Collins

Student Support Materials for

AQA

A2 BIOLOGY
SPECIFICATION (B)

Module 4: Energy, Control and Continuity

Mike Boyle
Series editor: Keith Hirst

This booklet has been designed to support the AQA Biology A2 specification. It contains some material which has been added in order to clarify the specification. The examination will be limited to material set out in the specification document.

Published by HarperCollins*Publishers* Limited
77–85 Fulham Palace Road
Hammersmith
London W6 8JB

10 9 8
ISBN-13 978-0-00-712416-9
ISBN-10 0-00-712416-3

British Library Cataloguing in Publication Data
A catalogue record for this publication is available from the British Library

Editor: Kathryn Senior
Front cover designed by Chi Leung
Design by Barking Dog Art, Gloucestershire
Printed and bound by Martins the Printers, Berwick Upon Tweed

The publisher wishes to thank the Assessment and Qualifications Alliance for permission to reproduce the examination questions.

Other useful texts

Full colour textbooks

Collins Advanced Modular Sciences: Biology AS
Collins Advanced Science: Biology

Student Support Booklets

AQA (B) Biology: 1 Core Principles
AQA (B) Biology: 2 Genes and Genetic Engineering
AQA (B) Biology: 3 Physiology and Transport
AQA (B) Biology: 5 Environment

To the student

What books do I need to study this course?

You will probably use a range of resources during your course. Some will be produced by the centre where you are studying, some by a commercial publisher and others may be borrowed from libraries or study centres. Different resources have different uses – but remember, owning a book is not enough – it must be *used.*

What does this booklet cover?

This *Student Support Booklet* covers the content you need to know and understand to pass the module test for AQA Biology A2 *Module 4: Energy, Control and Continuity.* It is very concise and you will need to study it carefully to make sure you can remember all of the material.

How can I remember all this material?

Reading the booklet is an essential first step – but reading by itself is not a good way to get stuff into your memory. If you have bought the booklet and can write on it, you could try the following techniques to help you to memorise the material:

- underline or highlight the most important words in every paragraph
- underline or highlight scientific jargon – write a note of the meaning in the margin if you are unsure
- remember the number of items in a list – then you can tell if you have forgotten one when you try to remember it later
- tick sections when you are sure you know them – and then concentrate on the sections you do not yet know.

How can I check my progress?

The module test at the end is a useful check on your progress – you may want to wait until you have nearly completed the module and use it as a mock exam or try questions one by one as you progress. The answers show you how much you need to do to get the marks.

What if I get stuck?

The colour textbook *Collins Advanced Modular Sciences: Biology A2* is designed to support your A2 course. It provides more explanation than this booklet. It may help you to make progress if you get stuck.

Any other good advice?

- You will not learn well if you are tired or stressed. Set aside time for work (and play!) and try to stick to it.
- Don't leave everything until the last minute – whatever your friends may tell you it doesn't work.
- You are most effective if you work hard for shorter periods of time and then take a (short!) break. 30 minutes of work followed by a five or ten minute break is a useful pattern. Then get back to work.
- Some people work better in the morning, some in the evening. Find out which works better for you and do that whenever possible.
- Do not suffer in silence – ask friends and your teacher for help.
- Stay calm, enjoy it and ... good luck!

The main text gives a very concise explanation of the ideas in your course. You must study all of it – none is spare or not needed.

The examiner's notes are always useful – make sure you read them because they will help with your module test.

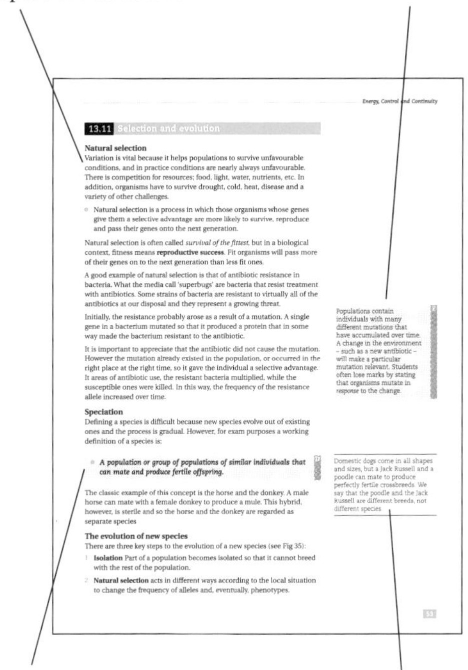

Energy, Control and Continuity

13.11 Selection and evolution

Natural selection

Variation is vital because it helps populations to survive unfavourable conditions, and in practice conditions are nearly always unfavourable. There is competition for resources; food, light, water, nutrients, etc. In addition, organisms have to survive drought, cold, heat, disease and a variety of other challenges.

- Natural selection is a process in which those organisms whose genes give them a selective advantage are more likely to survive, reproduce and pass their genes onto the next generation.

Natural selection is often called *survival of the fittest*, but in a biological context, fitness means **reproductive success**. Fit organisms will pass more of their genes on to the next generation than less fit ones.

A good example of natural selection is that of antibiotic resistance in bacteria. What the media call 'superbugs' are bacteria that resist treatment with antibiotics. Some strains of bacteria are resistant to virtually all of the antibiotics at our disposal and they represent a growing threat.

Initially, the resistance probably arose as a result of a mutation. A single gene in a bacterium mutated so that it produced a protein that in some way made the bacterium resistant to the antibiotic.

It is important to appreciate that the antibiotic did not cause the mutation. However the mutation already existed in the population, or occurred in the right place at the right time, so it gave the individual a selective advantage. It areas of antibiotic use, the resistant bacteria multiplied, while the susceptible ones were killed. In this way, the frequency of the resistance allele increased over time.

Populations contain individuals with many different mutations that have accumulated over time. A change in the environment – such as a new antibiotic – will make a particular mutation relevant. Students often lose marks by stating that organisms mutate in response to the change.

Speciation

Defining a species is difficult because new species evolve out of existing ones and the process is gradual. However, for exam purposes a working definition of a species is:

- ***A population or group of populations of similar individuals that can mate and produce fertile offspring.***

Domestic dogs come in all shapes and sizes, but a Jack Russell and a poodle can mate to produce perfectly fertile crossbreeds. We say that the poodle and the Jack Russell are different breeds, not different species.

The classic example of this concept is the horse and the donkey. A male horse can mate with a female donkey to produce a mule. This hybrid, however, is sterile and so the horse and the donkey are regarded as separate species

The evolution of new species

There are three key steps to the evolution of a new species (see Fig 35):

1. **Isolation** Part of a population becomes isolated so that it cannot breed with the rest of the population.
2. **Natural selection** acts in different ways according to the local situation to change the frequency of alleles and, eventually, phenotypes.

53

There are rigorous definitions of the main terms used in your examination – memorise these exactly.

Further explanation references give a little extra detail, or directs you to other texts if you need more help or need to read around a topic.

Section	Contents	Revision complete
13.1 Energy supply	• Relationship between respiration and photosynthesis. • ATP	
13.2 Photosynthesis	• Light dependent reaction • Light independent reaction • Chloroplast structure	
13.3 Respiration	• Glycolysis • Link reaction • Krebs cycle • Electron transport chain • Mitochondrial structure	
13.4 Survival and coordination	• Stimulus and response • Reflex arc • Hormone action	
13.5 Homeostasis	• Negative feedback • Temperature • Blood sugar • Waste removal • Water Potential	
13.6 Nervous coordination	• The eye • The nerve impulse • Synapses • Effects of drugs	
13.7 Analysis and integration	• The brain • The autonomic nervous system	
13.8 Muscles	• Antagonistic muscle action • Muscle structure • Sliding filament hypothesis	
13.9 Inheritance	• Genotype • Meiosis and fertilisation • Sex determination • Mono- and dihybrid inheritance	
13.10 Variation	• Types of variation • Causes of variation	
13.11 Selection and evolution	• Natural selection • Speciation	
13.12 Classification	• Principles of taxonomy • The five kingdoms	

13.1 Energy supply

The relationship between photosynthesis and respiration

From GCSE, you should remember the basic equations for these processes.

Photosynthesis:

Carbon dioxide and water $\xrightarrow[\text{chlorophyll}]{\text{light}}$ glucose and oxygen

Respiration:

Glucose and oxygen → carbon dioxide and water + energy

These two processes are effectively the reverse of each other. Photosynthesis traps energy in organic molecules and this is released in respiration. Overall, photosynthesis is the reduction of carbon dioxide to form organic molecules, while respiration is the oxidation of organic molecules to form carbon dioxide.

> **E** Be careful how you use the term 'energy'. The process of respiration *releases* the energy in organic molecules, and *transfers* it to ATP. Candidates often lose marks when they state that respiration *makes* energy. Energy cannot be made or destroyed.

What is reduction?

In reduction reactions, substances gain electrons. Electrons contain energy, so a substance that is reduced gains energy. When carbon dioxide is reduced to form carbohydrates such as glucose, the carbon dioxide molecule gains energy.

What is oxidation?

The opposite of reduction; oxidation is a reaction in which substances lose electrons and therefore lose energy. In respiration, organic molecules such as glucose have the electrons stripped from them, and the energy released is used to make ATP.

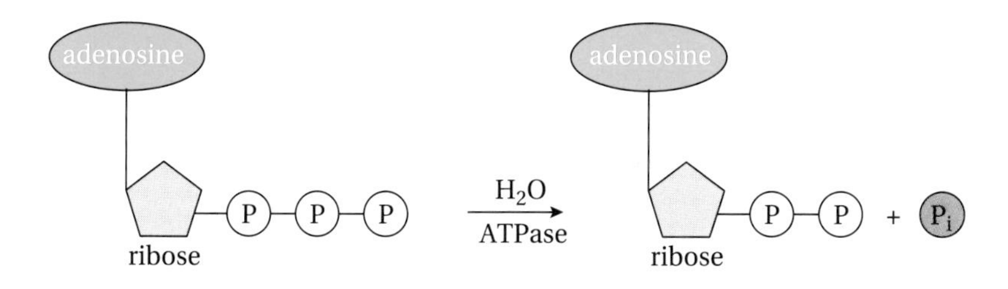

ATP ⟶ **ADP** + $\mathbf{P_i}$ + **energy**
(inorganic phosphate)

Fig 1
The hydrolysis (splitting) of ATP. This reaction releases energy that can be used to drive energy requiring reactions such as muscular contraction

Facts about ATP

- ATP stands for adenosine tri-phosphate (Fig 1). It can be **hydrolysed** (broken down) into adenosine di-phosphate (ADP) and free inorganic phosphate (P_i).
- Splitting ATP provides readily available energy in small, usable amounts for the wide variety of energy-requiring reactions that occur in cells
- ATP is an intermediate energy source within cells. It carries energy from the mitochondria, where respiration takes place, to areas of the cell where energy is needed.

- ATP is a relatively small molecule that can diffuse around the cell quickly.
- In respiration, the energy contained in organic compounds is released in a series of steps. This energy is used to make ATP from ADP and phosphate. Respiration can therefore be thought of as a process that replenishes ATP stocks, making sure that it is made as fast as it is used.
- The ATP, ADP and P_i in a cell are sometimes known as the 'phosphate battery'. If all of the phosphate were to exist as ATP, the battery would be 'fully charged'. However, ATP is an unstable molecule, and is constantly being broken down and re-synthesised in a cycle within the cell.

E ATP and energy are not the same. You cannot say that ATP is energy; ATP is a molecule that releases energy when it is hydrolysed to form ADP and phosphate.

13.2 Photosynthesis

Learning photosynthesis

Photosynthesis is often seen as a difficult topic, but you can make sense of it all if you approach it in the right way. Firstly, you need an overview; this provides a framework that you can fit the details onto. The basis of this framework is the division of photosynthesis into two easy steps.

1. **The light-dependent reaction**

   Light hits chlorophyll, which then emits two high-energy electrons. These electrons pass through a series of electron transfer reactions that make ATP and NADPH. The electrons in chlorophyll are replaced when water is split, a process that also produces oxygen as a by-product.

2. **The light-independent reaction**

   ATP and NADPH are used in a series of reactions known as the Calvin cycle. Overall, these reactions reduce carbon dioxide into glucose.

   Both steps take place in the chloroplast (see Fig 2), but it is important to remember that different reactions happen in different parts of this important organelle.

NADPH is similar to NADH, and performs basically the same electron-carrying function. It is useful to remember 'P for photosynthesis'; NADH in respiration, NADPH in photosynthesis. (NB: in reality, the P stands for phosphate.)

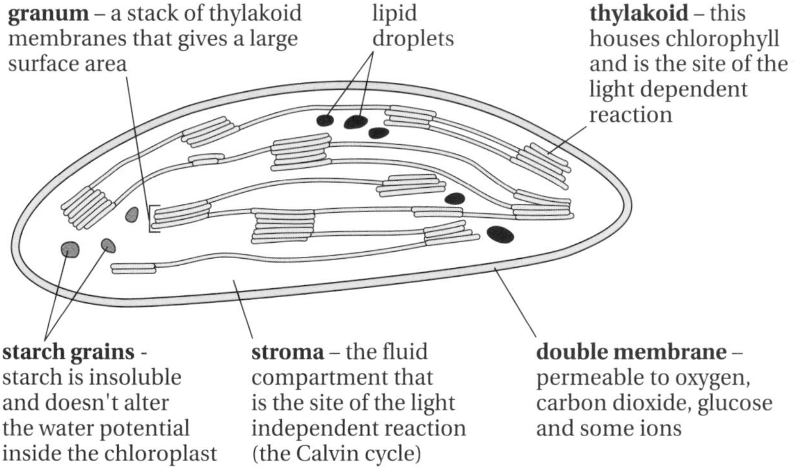

granum – a stack of thylakoid membranes that gives a large surface area

lipid droplets

thylakoid – this houses chlorophyll and is the site of the light dependent reaction

starch grains - starch is insoluble and doesn't alter the water potential inside the chloroplast

stroma – the fluid compartment that is the site of the light independent reaction (the Calvin cycle)

double membrane – permeable to oxygen, carbon dioxide, glucose and some ions

Fig 2
Chloroplasts are organelles that house all the enzymes and other substances needed for photosynthesis. They take the form of biconvex discs, rather like *Smarties*™ (but don't call them that in the exam)

The light dependent reaction in more detail

- This process happens on the **thylakoids**, membranes that contain the chlorophyll molecules.
- When a photon of light hits a molecule of chlorophyll, the pigment becomes excited and emits two high-energy electrons.
- These electrons pass down an electron transport system that synthesises ATP – this is called **photophosphorylation**.
- Energy from the excited electrons is also used to split water; this is **photolysis**. Photolysis produces electrons to replace those lost by the chlorophyll and protons (hydrogen ions) and oxygen are made as by products.

- Electrons are used to make NADPH (also called reduced NADP).
- ATP and NADPH are essential for the light independent reaction, and oxygen is released into the atmosphere

The light independent reaction in more detail

- This process happens in the **stroma**, the fluid in the centre of the chloroplast.
- It involves the reduction of carbon dioxide by a series of reactions known as the **Calvin cycle**.
- ATP and NADPH are used to reduce the carbon dioxide. Supplies of these compounds depend on light, so, when it gets dark, the light independent reaction finishes soon after the light-dependent reaction.
- The essential stages of the Calvin cycle are shown in Fig 3.

E A common mistake is to state that starch in a chloroplast provides extra energy for photosynthesis.

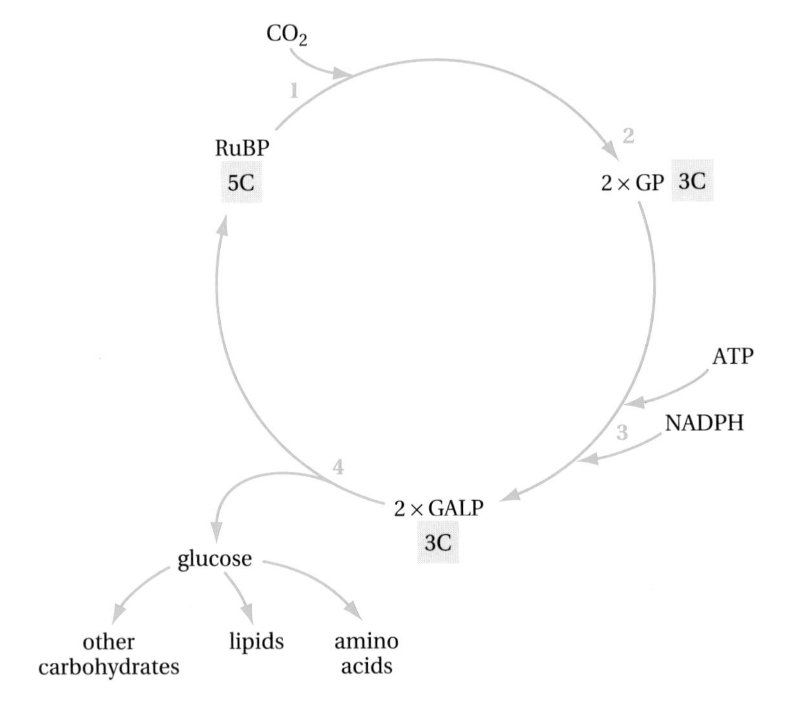

Fig 3
The Calvin cycle

1. Carbon dioxide combines with the 5-carbon compound **ribulose bisphosphate (RuBP)**. This reaction is catalysed by the enzyme **Rubisco** (ribulose bisphosphate carboxylase).
2. This produces a highly unstable 6-carbon compound that immediately splits into two molecules of **glycerate 3-phosphate (GP)**.
3. ATP and NADPH are used to reduce the GP into **glyceraldehyde 3-phosphate (GALP)**.
4. Some of the glyceraldehyde 3-phosphate is used to make carbohydrate, but most of it is used to make more RuBP to continue the cycle. For every glucose molecule produced, five molecules are RuBP are re-synthesised.

13.3 Respiration

Respiration is the release of energy from organic molecules. In this section, we study the breakdown of glucose. This is the main fuel for respiration in humans, but most other organic compounds can also be respired. For example, we respire fats when our carbohydrate stores run low, and many carnivorous animals such as cats get much of their energy by respiring amino acids from their high-protein diet. In order to understand respiration, you need to know about the co-enzyme NAD.

Respiration is an energy releasing process that takes place in virtually *all* living cells *all* of the time.

What is NAD?

NAD stands for Nicotinamide Adenine Dinucleotide, although you will not be asked to remember the full name. NAD is a coenzyme and its key feature is that it *carries electrons.*

$$NAD^+ + e^- \rightarrow NADH$$

Coenzyme + electron → reduced coenzyme

Respiration is basically a series of oxidation reactions that remove electrons from the original organic molecule. These electrons are picked up by NAD, forming NADH, also called **reduced NAD**. So whenever you see NADH, think 'an electron is being carried'.

NADH carries electrons into the electron transport system, where they are used to make large amounts of ATP. $FADH_2$, made in the Krebs cycle, is similar to NADH and does exactly the same electron-carrying function.

Glycolysis and the Krebs cycle

This title from the specification refers to just two processes, but in practice full aerobic respiration consists of four processes. These are shown in Fig 4 and are described below:

1. **Glycolysis**, in which one molecule of glucose is split into two molecules of **pyruvate**.
2. **The link reaction**, in which pyruvate is converted into **acetate**, which then combines with **co-enzyme A** to become **acetyl coenzyme A**.
3. **The Krebs cycle**, in which electrons are removed from the acetyl co-A.
4. **The electron transport system**, in which the energy in the electrons is used to make large amounts of ATP.

Fig 4
An overview of respiration showing glycolysis, the link reaction, the Krebs cycle and the electron transport chain. Glycolysis occurs in the cytosol of the cell, not inside the mitochondria. The link reaction and the Krebs cycle take place in the mitochondrial matrix, while the electron transport chain happens inside the thylakoid, across the inner mitochondrial membrane

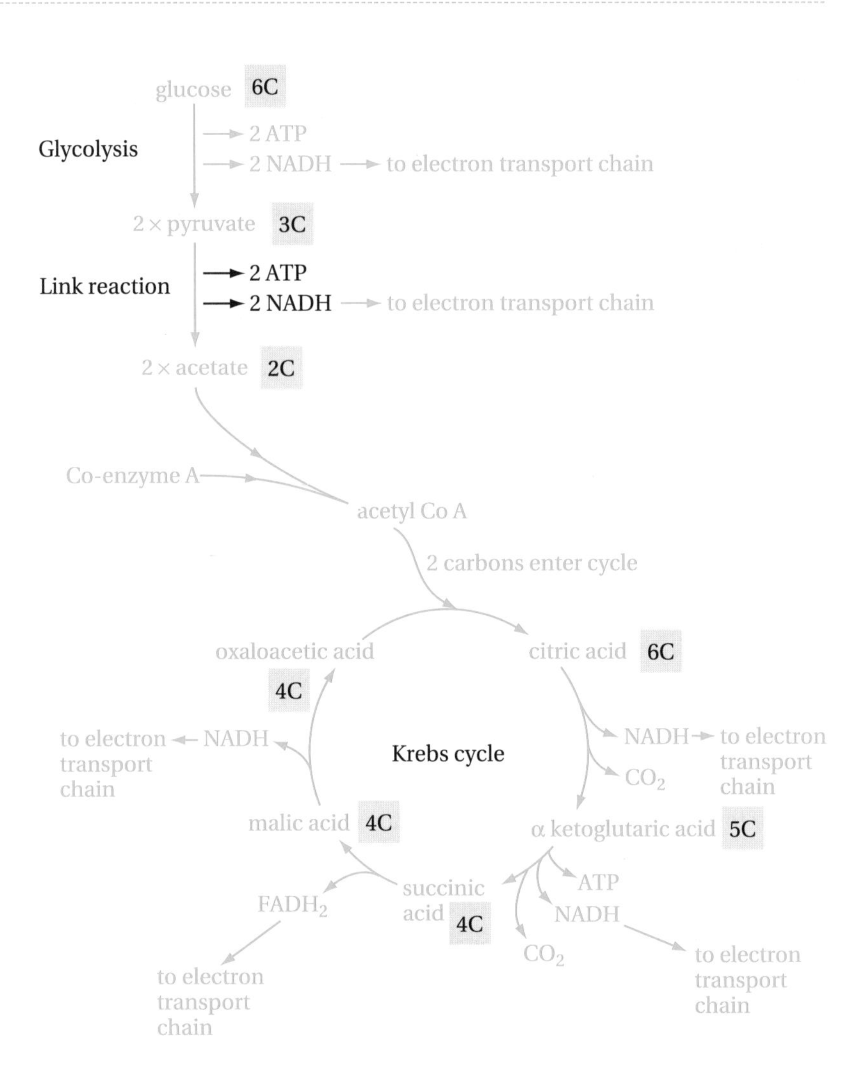

Electron transport chain

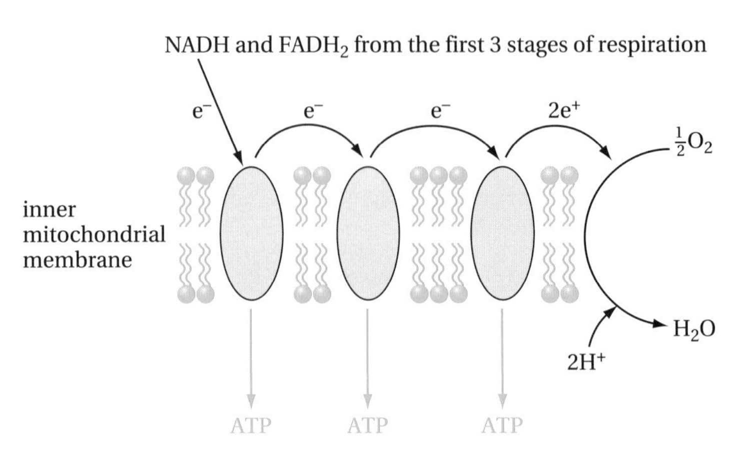

Respiration in more detail

The location of each reaction in the mitochondria is shown in Fig 5.

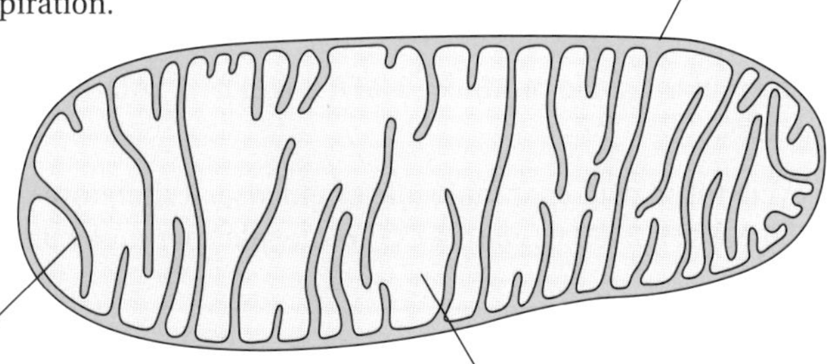

Fig 5
Structure and function of the mitochondrion

Glycolysis

- The word glycolysis means 'sugar splitting'.
- In glycolysis, one molecule of glucose is oxidised to produce two molecules of pyruvate (a 3 carbon compound).
- The reaction yields 2 molecules of ATP and 2 molecules of NADH for each glucose molecule that is split.
- Glycolysis takes place in the **cytosol** – the fluid part of the cytoplasm, *not* in the mitochondria.
- The process does not require oxygen. Anaerobic respiration is basically just glycolysis.

If respiration is aerobic i.e. if oxygen is available, pyruvate formed by glycolysis passes into the mitochondria. In anaerobic respiration, pyruvate is converted into lactate (or ethanol and carbon dioxide in plants and fungi). This reaction also oxidises NAD and so allows glycolysis to continue. However, when there is no oxygen available, no other stages of respiration occur.

Respiration and photosynthesis both consist of a series of steps. Each step is a relatively simple chemical reaction that is controlled by a specific enzyme.

The link reaction

So named because it links the anaerobic and aerobic parts of respiration.

- This reaction is also sometimes called pyruvate oxidation.
- In the link reaction, pyruvate is used to produce acetate and carbon dioxide. The acetate is picked up by coenzyme A, forming acetylcoenzyme A.
- No ATP is produced during the link reaction, but 2 molecules of NADH are formed.
- The link reaction takes place in the **matrix** (inner fluid) of the mitochondria.

The Krebs cycle

- This cycle is a series of reactions that oxidise what is left of the glucose after it has passed through glycolysis and the link reaction. By now, the glucose has been converted to acetate. In the Krebs cycle electrons are stripped from the acetate, creating large amounts of the electron carriers NADH and FADH.
- The reactions take place in the **matrix** of the mitochondria.
- Each turn of the cycle produces 1 ATP, 3 NADH,1 $FADH_2$ and 2 CO_2 molecules.
- The cycle turns twice per molecule of glucose. It therefore produces 2 ATP, 6 NADH, 2 $FADH_2$ and 4 CO_2 molecules per molecule of glucose.

How much ATP is made per glucose?

In ideal conditions the theoretical maximum is 38. Here we see why.

There are two ways of making ATP:

1. By **substrate level phosphorylation**. This is the method of ATP production in glycolysis and the Krebs cycle. Both processes provide 2 ATPs per glucose, giving a total of 4 ATPs by this method.
2. By **oxidative phosphorylation**. Here, ATP is made using energy released in the electron transfer system. In effect, it is ATP made by 'cashing in' the energy in the electrons carried by NADH and $FADH_2$. The other 34 molecules of ATP made during respiration per molecule of glucose are made by this method, as you can see below.

Table 1 shows the total ATP and NADH production from the first three parts of respiration, per molecule of glucose.

Table 1
ATP and NADH made during glycolysis, the link reaction and the Krebs cycle

Process	ATP made	NADH made	$FADH_2$ made
Glycolysis	2	2	0
Link reaction	0	2	0
Krebs cycle	2	6	2
Totals	4	10	2

When fed into the electron transport chain, 3 ATP molecules are made per NADH, and 2 per $FADH_2$

This gives us…

From NADH $10 \times 3 = 30$ ATP

From FADH2 $2 \times 2 = 4$ ATP

This gives a total of 34 molecules of ATP made by oxidative phosphorylation. When we add these to the original 4 produced by glycolysis and the Krebs cycle, we get the 38 ATPs produced per molecule of glucose during aerobic respiration.

Note the similarities between chloroplasts and mitochondria (page 7 and 11, respectively). Both are about the same size (roughly 10 μm across) and both have an internal membrane system giving a large surface area for reactions.

The electron transport chain

- This consists of a series of proteins on the inner mitochondrial membrane, which is folded into **cristae** to provide a large surface area for the process.
- Electrons are delivered to the electron transport system by NADH and $FADH_2$.
- The electrons pass from one protein to the next along the chain.
- Each electron transfer is an oxidation/reduction reaction that releases energy.
- This energy is used to make ATP.
- The by-products of this process are low energy electrons and protons. The protons combine with oxygen to form water.
- You are breathing at this moment because of the electron transport chain. You need to provide oxygen to mop up the electrons, and also to remove the accumulated carbon dioxide from your lungs.

13.4 Survival and coordination

Stimulus and response

A feature of all organisms is the ability to detect and respond to changes in their surroundings. In this section we study coordination in mammals, though the basic principles apply to many other organisms.

First, some definitions:

D

- **Stimulus** – *a change in an organism's environment (internal or external) that can be detected by receptor cells.*
- **Receptor** – *a specialised cell that detects a stimulus and initiates a nerve impulse.*
- **Sensory neurone** – *a nerve cell that carries impulses from receptor to the central nervous system.*
- **Central Nervous System** (CNS) – *the brain and spinal cord. The CNS processes incoming information and produces a response, often based on previous experience.*
- **Motor neurone** – *a nerve cell that carries impulses from the CNS to the effectors.*
- **Effectors** - *organs that bring about a response, usually muscles or glands.*

Generally, we respond to *changes* in our environment while constant stimuli, such as humming of a florescent light, are ignored. Many changes in the environment are not stimuli because the organism does not have the appropriate receptor to detect them. For example, very high-pitched sounds are not a stimuli in humans because they are beyond the range of our hearing.

Fig 6
The reflex arc represents the most direct way of connecting up a sense organ to an effector, producing the fastest possible response

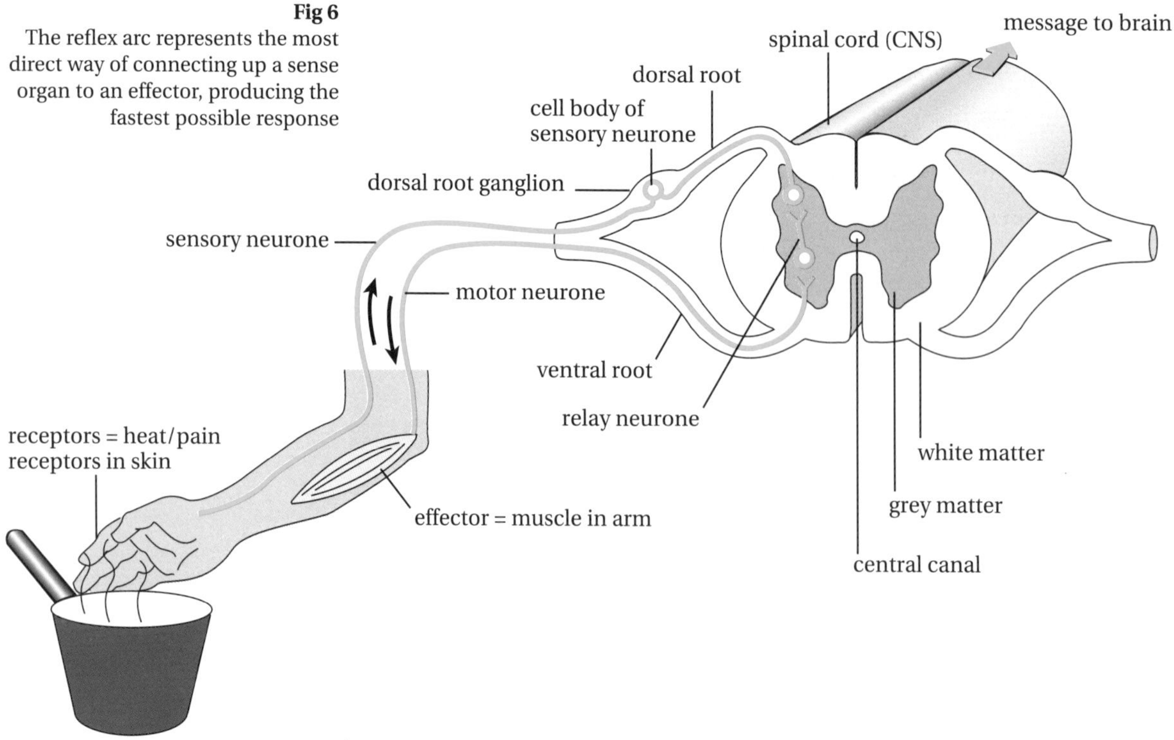

The reflex arc

The **reflex arc** is the simplest example of coordination. A particular stimulus leads to a fixed response – this is very rapid and can't be controlled because the nerve impulses involved do not need to pass through the conscious parts of the brain for the response to take place.

Fig 6 shows an example of a reflex that avoids danger and minimises damage. Other reflexes, such as the knee jerk, are postural reflexes; one of the many mechanisms we have to maintain our position and body control without having to constantly think about fine adjustments. Most reflexes are either postural or have evolved to avoid danger.

The knee jerk reaction is a three-neurone reflex arc. In this example the reflex is initiated by the stretch receptor is the **patellar tendon**. A tap in this tendon, just below the knee, causes the lower leg to kick forwards.

The three neurones involved are:

1. A **sensory neurone** that links the receptor with the spinal cord.
2. A short **relay neuron** that connects the incoming and outgoing neurones.
3. The **motor neurone** that transmits impulses to the effector.

Sensory information that the knee jerk has happened also passes up the spinal cord to the brain. The response happens well before the conscious brain has a chance to stop it, but you are aware that your leg has moved.

E

Reflex arcs that result from stimuli to the head - e.g. to make us blink - pass through the brain, not the spinal cord. However, they are still not under conscious control

Hormones and targets

Again, let's look at some definitions:

D

- *An **endocrine gland** is one that synthesises and secretes hormones. These glands have no ducts ('delivery tubes'). They release the hormone directly into the blood. Examples of endocrine glands include the pituitary, adrenal and thyroid glands.*
- *A **hormone** is a chemical produced by an endocrine gland. Hormones travel in the bloodstream and have an effect on particular target organs or cells.*

Generally, hormones are **peptides** (short chains of amino acids), proteins or lipids. Lipid hormones are called **steroids**. Hormones work by fitting into specific receptor proteins on the surface of target cells – only the target cells have the 'right' receptors (Fig 7). When the hormone binds to the receptor, this triggers a change in activity within the cell.

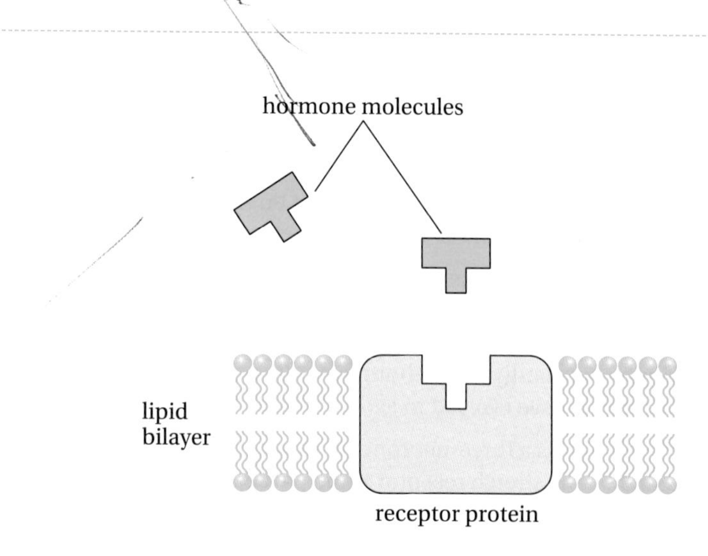

Fig 7
Hormone molecules fit exactly into receptors on the surface of their target cells

As an example, the female sex hormone FSH is secreted by the pituitary gland. FSH circulates to all parts of the body, but the ovaries are the target organs. Their cell membranes have FSH receptors. When FSH fits into the receptors, the ovarian cells respond by secreting the hormone oestrogen. In turn, oestrogen targets specific cells all over the body.

13.5 Homeostasis

Homeostasis is the ability of an organism to maintain its internal conditions within particular limits. In mammals, examples of homeostasis include:

- Maintaining the pH of blood and body fluids between 7.3 and 7.45.
- Maintaining the core body temperature at around 37 °C.
- Maintaining blood glucose levels between 4 and 11 millimoles per litre.

You can see from these examples that internal conditions are not absolutely constant, but are maintained within fairly narrow limits.

There are many other examples of homeostasis, such as the maintenance of all of the different hormone levels and the many components of blood plasma.

Negative feedback – the mechanism of homeostasis

Most examples of homeostasis involve negative feedback. When a factor changes, the homeostatic mechanism acts to reverse that change and bring things back to normal. Fig 8 shows a negative feedback loop and how it differs from positive feedback.

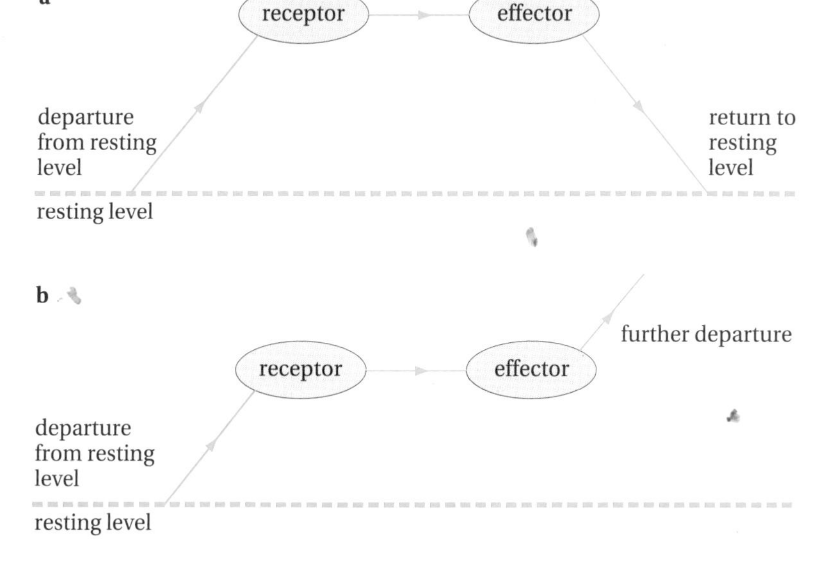

Fig 8
a Negative feedback is a mechanism that keeps levels constant.
b In contrast, positive feedback is a mechanism for bringing about change

When you discuss an example of homeostasis, it is important to consider all of the following:

1. What causes the level to change?
2. What detects the change?
3. How is the change reversed?

These three aspects are all included in the three examples of homeostasis that follow.

1 Regulation of body temperature

Regulating body temperature basically involves the internal generation of heat (produced as a by-product of metabolic reactions) and the control of its release into the environment. When we are too hot, we maximise our heat loss to the environment, and when we are too cold, we minimise it. We can also alter our metabolic rate to produce more or less heat in the first place.

Some definitions:

D

- ***Endotherm** means 'heat from within.' Mammals and birds are endotherms, they have the ability to control their core temperature regardless of the external temperature. Endotherms control their body temperature by physiological and behavioural means. Endotherms used to be described by the now out-dated term warm-blooded.*
- ***Ectotherm** means 'heat from the outside' and refers to animals that can only regulate their body temperature by behavioural means. In practice their body temperature is usually similar to that of the environment. Ectotherms used to be described by the now out-dated term cold-blooded.*

Behavioural thermoregulation in mammals

Although mammals and bird have physiological ways of controlling their core temperature, their behaviour still plays a large part. When we feel cold, for example, we change our posture and surroundings by, for example, folding our arms, putting on more clothes or by closing a window.

The temperature of the blood is detected by the **hypothalamus**, which is basically the body's thermostat. Inside the hypothalamus is the **thermoregulatory centre**, which has two parts; a **heat loss** centre and a **heat gain** centre. Fig 9 gives an overview of the mechanisms of thermoregulation.

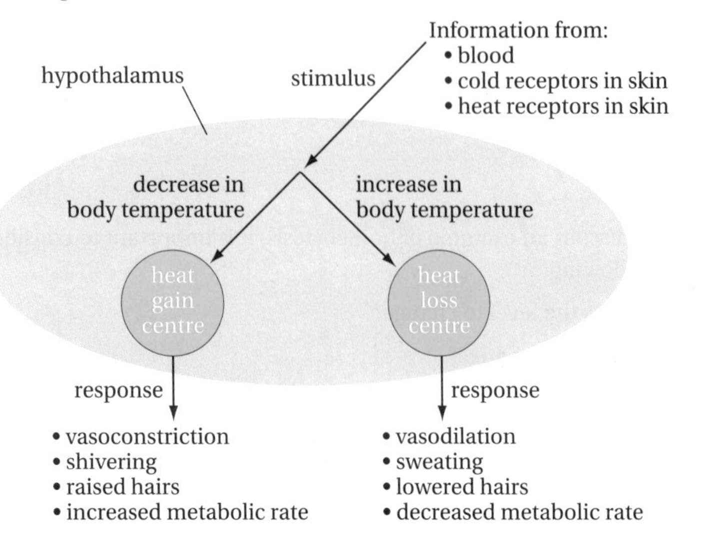

Fig 9
The role of the heat gain centre and heat loss centre of the hypothalamus in temperature regulation

The skin plays an important role in thermoregulation; its structure is shown in Fig 10.

We get too cold when the heat loss from our bodies exceeds what is generated, the temperature of the blood falls. This drop is detected by the hypothalamus, and signals pass from the heat gain centre to bring about the following responses:

1. **Shivering** – rapid contraction and relaxation of the muscles that creates heat. This heat is distributed to the rest of the body in the blood
2. **Vasoconstriction** – constriction of the muscular arteriole walls re-direct blood away from the skin, this keeping the warmer blood in the centre of the body (see Fig 11a).
3. **Making your hair stand on end** – hairs are raised by means of erector pili muscles. In hairy mammals this traps a layer of insulating air, but in humans its not so hair-raising – it just gives us goose pimples.
4. **Increasing basal metabolic rate** – in the short term, this is achieved by the hormone adrenaline and, in the long term, by secretion of the hormone thyroxine.

E Arterioles are fixed, they can constrict or dilate, but they *do not* 'move to the surface of the skin'.

We get too hot when we generate more heat than we lose. During exercise, for example, the temperature of the blood rises. In this case, the heat loss centre brings about:

1. **Sweating** – secretion of a salty solution from the millions of sweat (**sudorific**) glands that cover our bodies. Sweat cools the body as it evaporates (water has a high latent heat of evaporation, which means that water molecules take a lot of energy with them as they change state from liquid to vapour).
2. **Vasodilation** – the arterioles dilate, allowing more blood to flow through the capillaries of the skin. The heat in the blood is transferred to the sweat, and is lost from the body as the sweat evaporates (see Fig 11b).
3. **Lowering of hairs** as erector pili muscles relax.

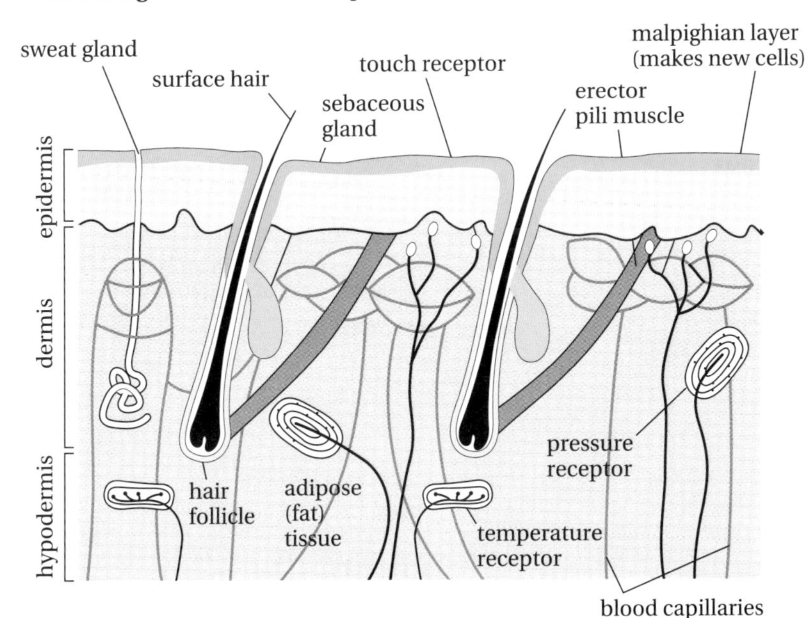

Fig 10
The structure of human skin

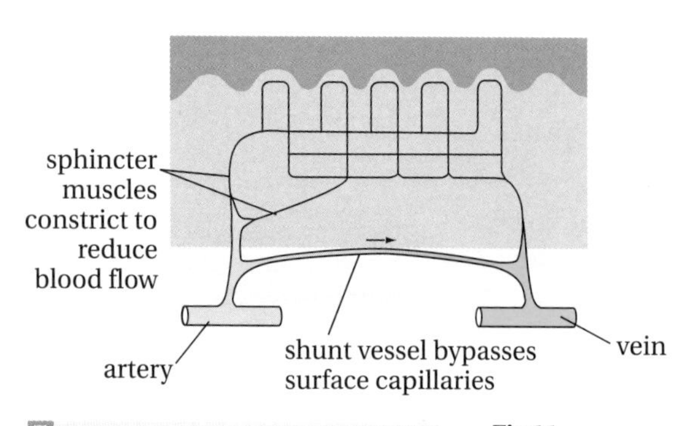

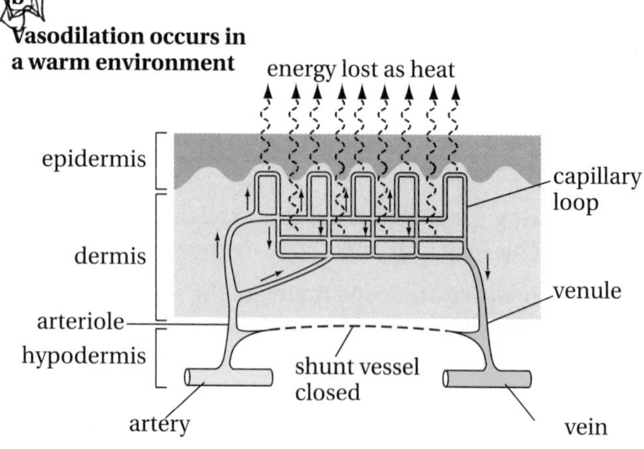

Fig 11
Blood flow to the skin can be altered by controlling the flow of blood in the arterioles leading to the surface capillaries

> **E** Be careful about the difference between 'heat' and 'temperature'. Strictly speaking, things don't contain heat – sweat does not 'take heat away' when it evaporates. It is more accurate to say that energy is transferred from the blood to the water molecules in sweat, the water molecules have more kinetic energy and evaporate.

2 Regulation of blood glucose levels

The **pancreas** is the key organ in the control of blood glucose levels because it makes the hormones **insulin** and **glucagon**. Over 90 % of the pancreas is dedicated to making pancreatic juice, but there are small patches of cells, the **islets of Langerhans**, that make and secrete hormones (see Fig 12). The islets consist of two types of cells; α (alpha) cells and β (beta) cells. The beta cells secrete **insulin** and the alpha cells secrete **glucagon**. These hormones are **antagonistic**; they have opposing effects.

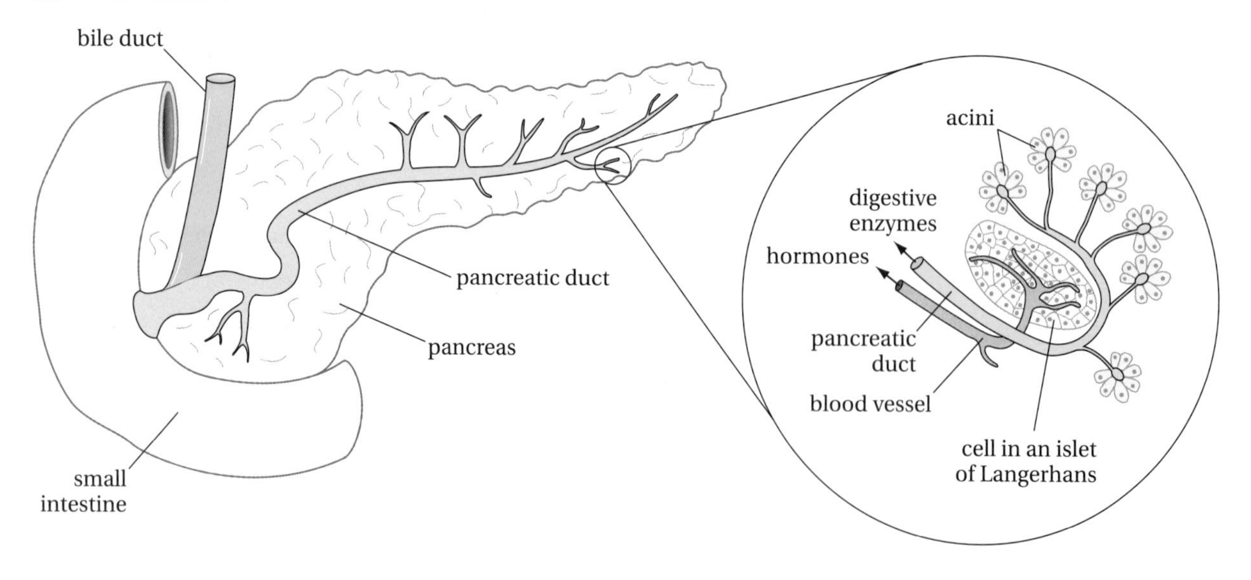

Fig 12
The major part of the pancreas makes a juice containing digestive enzymes, but small patches of cells, the islets of Langerhans produce the hormones insulin and glucagon

These definitions are useful for this topic:

- ***Glycogenesis** – the production of glycogen.*
- ***Glycogenolysis** – the breakdown of glycogen.*
- ***Gluconeogenesis** – the production of glucose from non-carbohydrate sources i.e. lipid or protein. This happens during fasting/dieting/starvation, when glucose and glycogen levels are low.*

D

The terms glycogenesis, glycogenolysis and gluconeogenesis look similar and it is easy to confuse them. Look at the key parts of the words: **genesis** = creation, **lysis** = splitting, **neo** = new.

If blood sugar levels are too high...

Blood glucose levels usually rise for a few hours following a meal, as the sugars and starches are digested into glucose and absorbed into the blood. The high levels are detected by the beta cells themselves, which respond by secreting insulin. This hormone travels in the blood and fits into specific receptor proteins on membranes of cells throughout the body, but notably those of the liver.

Blood sugar high
beta cells. secrete insulin
fit in receptor on liver.
glucose channels open.
influx of glucose into liver
→ glycogen.

When insulin fits into the receptor proteins, extra glucose channels open, allowing glucose to pass from the blood into the cells where it can be metabolised (see Fig 13). Insulin also activates enzyme pathways that convert glucose into glycogen, protein or lipid.

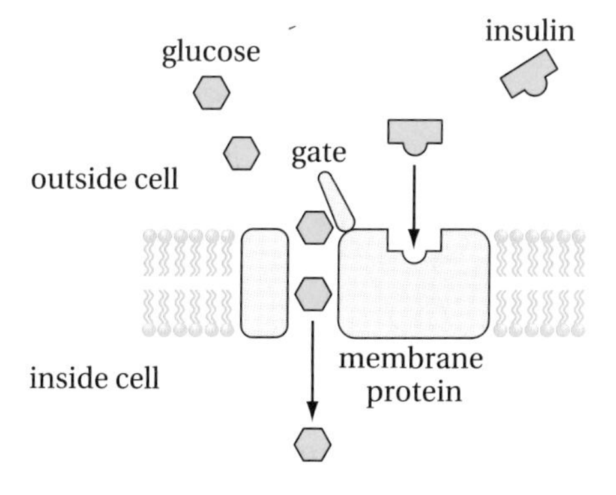

Fig 13
Insulin binding opens gates that allow glucose molecules into the cell

Diabetes

Insulin dependent diabetics cannot make insulin and so glucose cannot pass from blood into cells when needed. So blood glucose levels can rise while the cells are starved of fuel. Symptoms of diabetes include:

- **Thirst** – the high glucose levels decrease the water potential of the blood, stimulating the sensation of thirst.
- **Glucose in the urine** – the kidneys normally reabsorb all glucose, but when blood glucose levels are high they are unable to do so.
- **Weight loss** – when starved of their main fuel, cells respire other fuels such as lipid.
- **Breath smells of ketones** (a fruity smell) – keto acids are a by-product of lipid metabolism.
- **Excessive urination** – a consequence of the increased drinking.

Insulin is a relatively small protein, consisting of 51 amino acids in two polypeptide chains.

3 Removal of metabolic waste by the kidneys

D

- *Excretion is the removal from the body of the waste products of metabolism, which would be toxic if allowed to accumulate.*

From this definition, you can see that exhaling (removing carbon dioxide) and sweating (water and salts) are forms of excretion, but the main process involves production of urine and the main organs of excretion are the kidneys (see Fig 15).

The body can store carbohydrate and lipid, but it cannot store protein. An average healthy adult human needs about 50 to 60 g of protein per day – about the size of a large egg. Any extra must be broken down and excreted.

Following digestion of protein in the gut, amino acids are absorbed into the blood and travel to the liver in the hepatic portal vein. The liver can do one of four things to amino acids:

1. Use them to synthesise proteins, such as fibrinogen.
2. Release them unchanged into the circulation.
3. Break them down – this is **deamination**.
4. Change them into a different amino acid – this is **transamination**.

Deamination

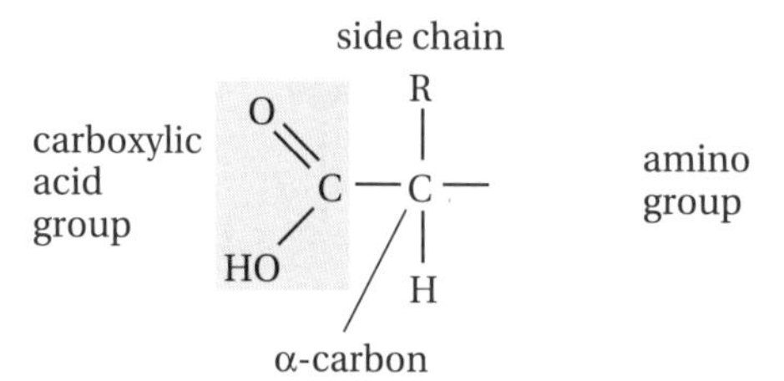

Fig 14
The basic structure of an amino acid

The process of deamination occurs in the cells of the liver, and involves the removal of the amino group. The products are ammonia (NH_3) and an organic acid.

The acid is respired while the ammonia, which is highly toxic, is converted into the relatively harmless **urea** via a simple cycle of reactions known as the **ornithine cycle**.

As the equation below shows, urea consists of two molecules of urea combined with one molecule of carbon dioxide:

$$2NH_3 + CO_2 \rightarrow CO(NH_2)_2 + H_2O$$

The urea produced in the liver passes into the blood where it is removed and excreted by the kidneys.

The role of the kidneys

The kidneys have three functions:

1. Removal of metabolic waste, notably urea.
2. Control of the volume of body fluids.
3. Control of the water potential of body fluids (i.e the concentration of dissolved substances).

Fig 15

a Section through the kidney

nephron of kidney tubule
collecting duct
renal capsule
glomerulus
pyramid
loop of Henle
cortex
medulla
renal artery
pelvis
renal vein
glomeruli
ureter
waves of peristalsis force urine down to the bladder
arterioles lead into and out of a glomerulus

b The nephron and its blood supply

renal capsule
first convoluted tubule
glomerulus
second convoluted tubule
cortex
1
3
4
vein
artery
collecting duct
2
loop of Henle (ascending limb)
medulla
loop of Henle (descending limb)
pyramid in the medulla
opening to pelvis

c Fine structure of the renal capsule

wide afferent arteriole
narrow efferent arteriole
slit in capillary endothelium
filtrate in lumen of capsule
capillary
podocyte
red blood cell
basement membrane
glomerular capillaries
blood plasma
layer of capsule cells
podocyte
podocyte cell of renal capsule wall
filtrate in lumen of capsule

The walls of capillaries in the renal capsule are more permeable than those of normal capillaries: the cells have thin slits between them, through which all the constituents of the plasma can pass

The renal capsule is lined with unique cells called podocytes ('foot-cells'). These cells also form a network of slits

Between these two relatively coarse filters is the continuous basement membrane. This finer filter prevents the passage of larger molecules (mainly proteins), which remain in the blood

d Absorption of filtrate from first convoluted tubule

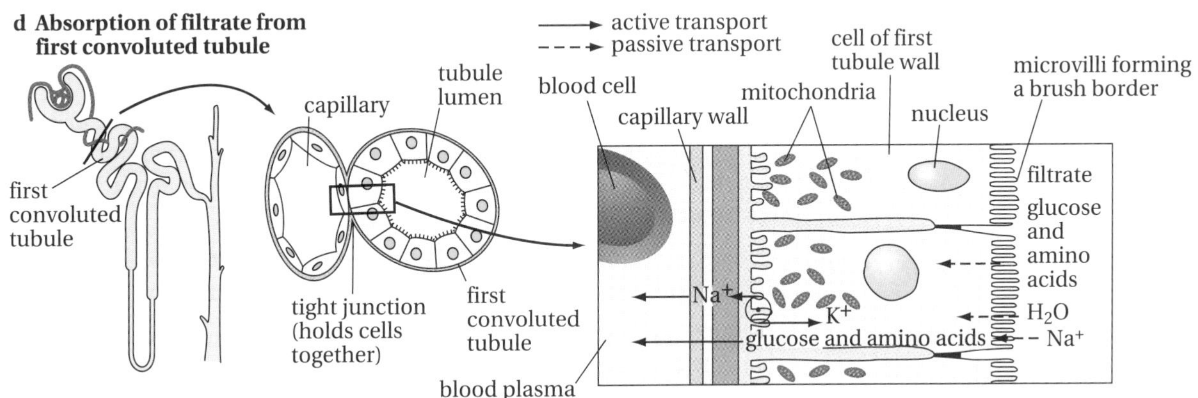

hydrostatic pressure.

Figs 15a and b show the overall structure of the kidney and the position of the kidney tubules, or **nephrons**. Each kidney contains over a million nephrons.

As an overview, kidney function can be divided into two key processes:

1. **Ultrafiltration**. Blood is filtered under pressure to produce filtrate; a substance identical to tissue fluid.
2. **Selective reabsorption and secretion**. A variety of active transport mechanisms secrete unwanted substances into the filtrate, while reabsorbing what is needed in to the blood. The end product is urine.

Ultrafiltration in the Bowman's capsule

Blood entering the kidney splits up into many small arterioles, each of which passes into a Bowman's capsule. The **afferent arteriole** (the one going in) is wider than the **efferent arteriole** leading blood away, and this further increases the **hydrostatic pressure** created by the left ventricle of the heart.

Blood is forced against a filter that consists of three layers (Fig 15c):

1. The **capillary endothelium**.
2. The **basement membrane** - a sheet of protein fibres that lies between the two layers of cells. It acts as a fine filter.
3. The **pedicels** (foot-like projections) on the **podocytes** lining the capsule.

Blood is forced against this filter, forming filtrate at the rate of about 120 cm^3 per minute. Clearly, we do not make this much urine, so it is obvious that the majority of this fluid is reabsorbed.

The two twisted portions of the nephron are known by several different names. The first one, the proximal (= close) convoluted tubule, sometimes just called the proximal tubule or the PCT. The distal (= distant) convoluted tubule is also the distal tubule or DCT.

Reabsorption in the proximal convoluted tubule

The blood that leaves the Bowman's capsule has lost much of its hydrostatic pressure but it still has a low water potential. This is due to the high concentration of plasma proteins that were too large to be filtered. Thus, when blood comes into contact with the filtrate again, water passes from the filtrate by osmosis, and many of the dissolved substances diffuse back into the blood.

You can see from the Fig 15d that the tubule cells show classic adaptations for active transport; a large surface area provided by microvilli, and many mitochondria to provide the necessary ATP. Active transport mechanisms secrete unwanted substances such as urea into the filtrate, while useful compounds, such as glucose and amino acids, are reabsorbed.

purpose of loop of Henle:
produce hypertonic urine
water - osmosis out.
- Sodium/chlorine diffuse into descending limb
sodium/chlorine actively pumped out.
low water potential in medulla.
water osmosis collecting duct.

Water reabsorption in the loop of Henle

The purpose of the loop of Henle is to allow us to make **hypertonic urine** i.e. urine with a lower water potential than the blood. This is important in land living animals as it allow us to excrete without losing too much precious water. The essential steps are (see also Fig 16):

1. Sodium and chloride ions diffuse into the filtrate in the descending limb, so the water potential decreases as it flows towards the apex.
2. As the filtrate flows up the ascending limb, sodium and chloride ions are actively pumped out.

3 The overall effect is to create a region of high salt concentration deep in the medulla, through which the collecting duct must pass. If the body needs to conserve water, specific water channels open, allowing water to pass out of the filtrate and so be conserved in the body.

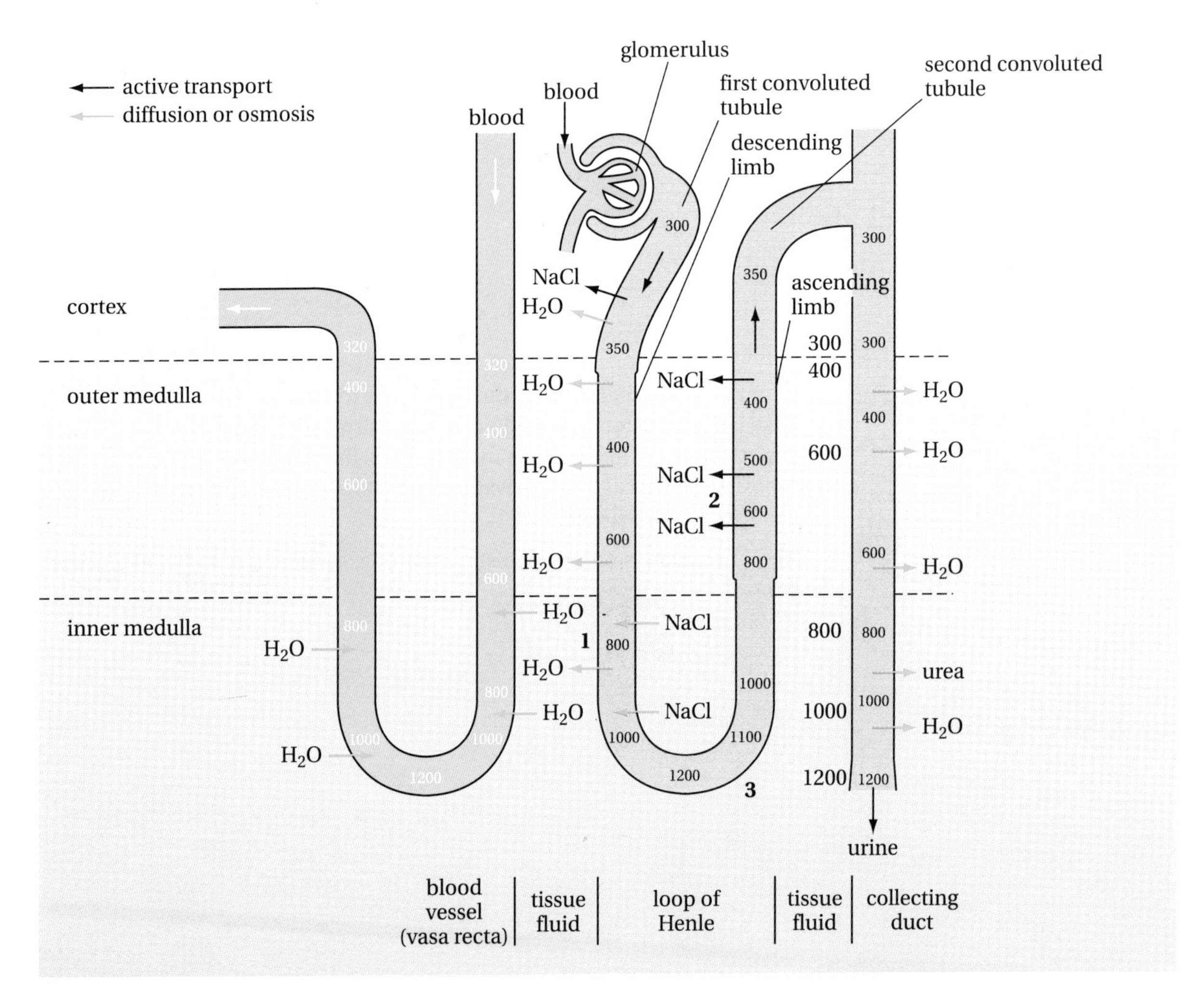

Fig 16
How the loop of Henle works. The bold numbers match the steps in the text while the numbers in the tubules and blood vessel refers to solute concentration

The distal tubule and the regulation of blood water potential

When the water potential of the blood gets too low, i.e. when we are beginning to dehydrate, the hypothalamus detects this and responds by stimulating the pituitary gland to secrete **anti-diuretic hormone (ADH)**. This hormone circulates in the blood and targets the cells of the distal tubule and collecting duct (Fig 17). Hormone molecules fit into receptor sites and open water channels, allowing water to pass out of the filtrate and re-enter the blood. In this way the body conserves water, while producing a smaller volume of more concentrated urine.

hypothalamus → pituitary gland → ADH → distal tubule/collecting duct → open water channels → water enters blood.

Fig 17
How ADH acts to produce hypertonic urine

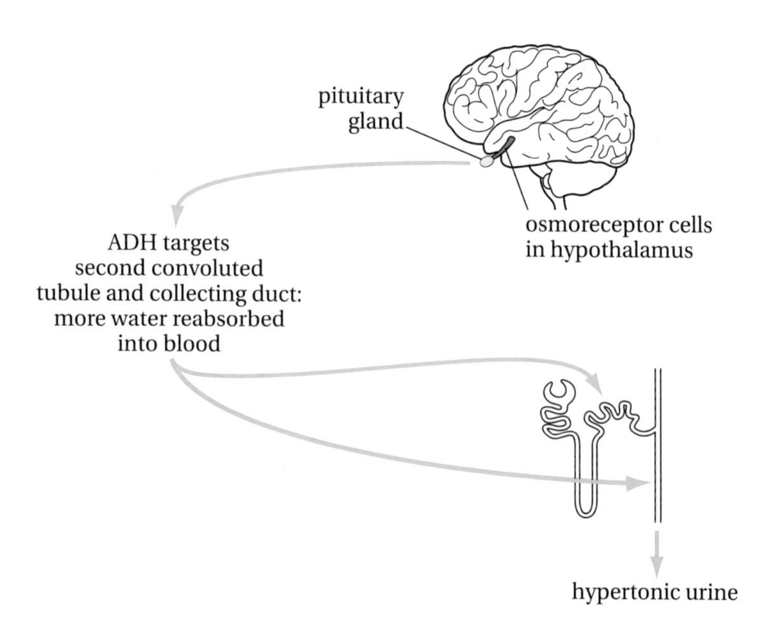

When the water potential of the blood gets too high – usually because we have consumed too much liquid – this is also detected by the hypothalamus. The response in this case is the inhibition of ADH production. This allows less water to be reabsorbed, resulting in larger volumes of dilute urine.

Note that there is no antagonistic hormone, as there is with blood glucose control.

Caffeine and alcohol are **diuretics** – they make the kidneys remove more water then the body needs. Many of the symptoms of hangovers are due to dehydration.

13.6 Nervous coordination

The mammalian eye

Fig 18 gives an overview of the mammalian eye. At this level you need to know about three separate aspects of eye function; the **iris**, **focusing** and the **formation of images by the retina**.

a

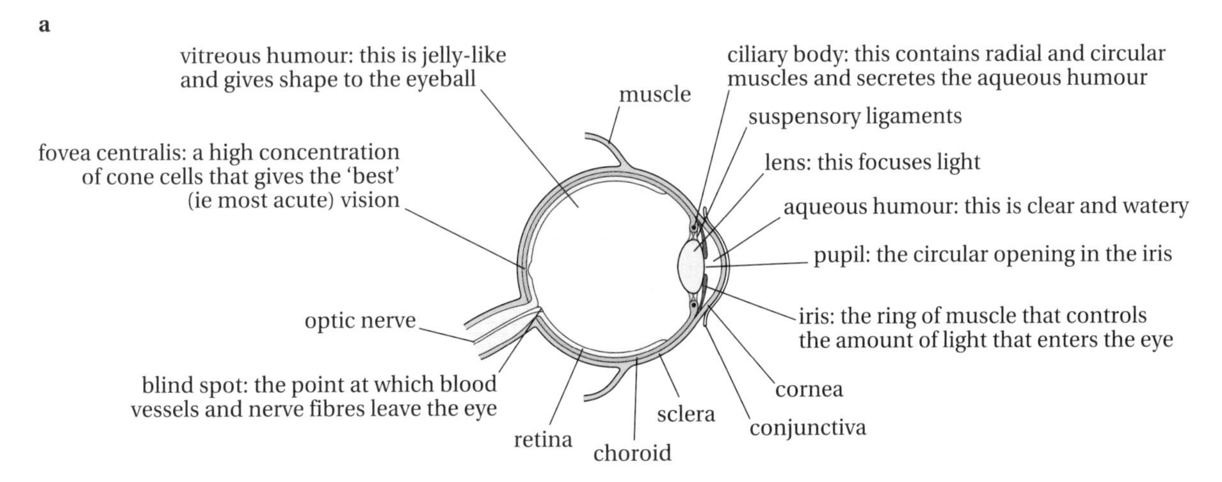

b

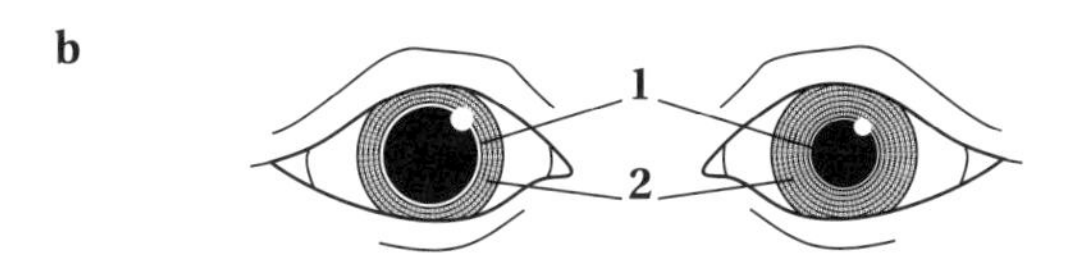

1 Circular muscle relaxed
2 Radial muscle contracted.
Pupil dilated (large)

1 Circular muscle contracted
2 Radial muscle relaxed.
Pupil constricted (small)

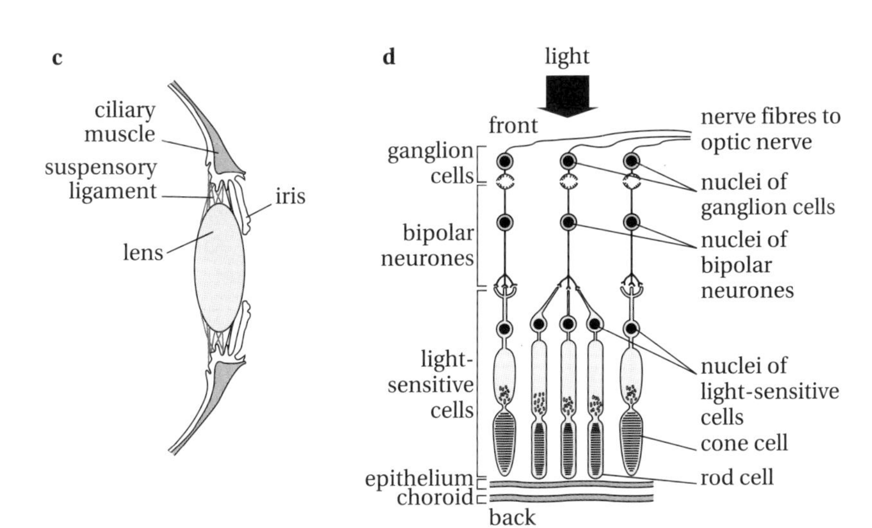

Fig 18
a Structure of the human eye
b How the circular and radial muscles allow us to see well in dim and bright light
c How the ciliary muscles and suspensory ligaments help with focusing
d Detailed structure of the retina

Fig 18a shows the basic structure of the eye.

The iris controls the amount of light entering the eye and so falling on the retina. This is important because the retina has an optimum light level at which it functions best. The iris contains two sets of antagonistic muscles. **Circular muscles** go around the iris, while **radial muscles** pass from the inside to the outside (Fig 18b)

In dim light, the radial muscles contact while the circular muscles relax, increasing the size of the pupil and allowing more light to reach the retina. In bright light the opposite happens so that the pupil constricts, reducing the amount of light that can get in.

Focusing

Focusing is also called **accommodation**.

E Candidates often state that when the ciliary muscle is relaxed, so is the lens. In reality, when the eye is relaxed, the lens is under tension and focused on distance.

All of the transparent structures in the eye, notably the cornea, have a refractive index; they deflect light by a specific amount. The lens can change its shape, giving it a *variable* refractive index. Thus the lens plays a central role in focusing.

When our eyes are closed, the ciliary muscles are relaxed. In this state, the suspensory ligaments are under tension and the lens is pulled flat. When we open our eyes we are focused on distant objects. To focus on a nearby object, the ciliary muscles contract, the tension on the suspensory ligaments lessens and the lens can assume its 'normal' rounded shape (Fig 18c).

The retina

The retina is a single layer of light sensitive cells – **rods** and **cones** – together with connective neurones on the back of the eye (Fig 18d). The overall function of the retina is to gather information about the incoming light and relay it to the brain via the **optic nerve**. The actual image is formulated in the brain, in the **visual cortex** (Fig 19).

Fig 19
The location of the visual cortex in the human brain

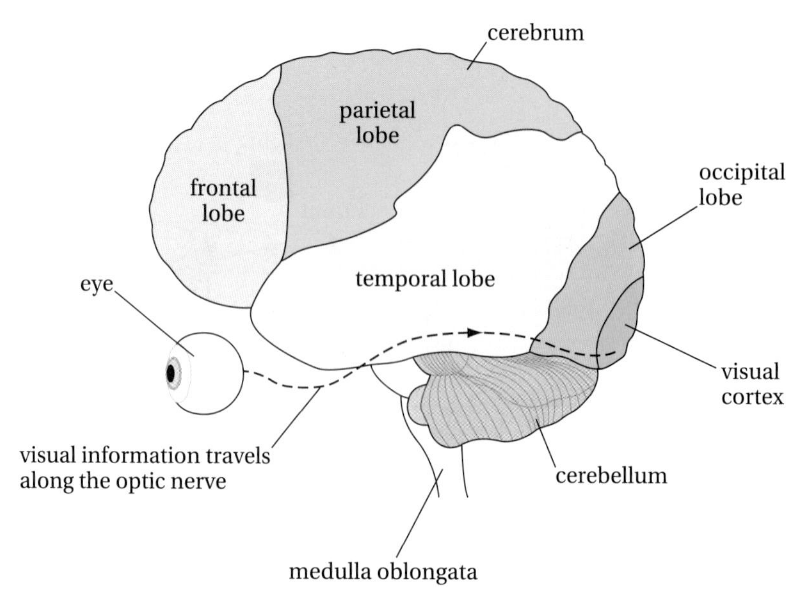

Feature	Rod cells	Cone cells
Shape	Rod-shaped outer segment	Cone-shaped outer segment
Connections	Many rods converge into one neurone	Only a single cone per neurone at centre of fovea
Visual acuity	Low	High
Visual pigments	Rhodopsin	Iodopsin
Numbers	120 million per retina	7 million per retina
Distribution	Found evenly all over retina	All over retina, but more concentrated towards fovea. The fovea itself consists just of cones
Sensitivity	Sensitive to low light intensity	Only functions in bright light
Overall function	Black and white vision in poor light	Seeing colour and detail in bright light

Table 2
The differences between rods and cones.

How a rod cell works

Rod cells (Fig 20) contain a light-sensitive pigment called **rhodopsin**, which consists of two components, **retinal**, made from vitamin A, and **opsin**, a protein. When light strikes rhodopsin, the molecule splits, producing free opsin. This acts an as enzyme, initiating a series of reactions that leads to hyper-polarisation of the rod cell membrane (the outside gets more negative). If this charge reaches a *threshold*, an action potential is initiated in the connecting neurones, and passes down the optic nerve to the brain.

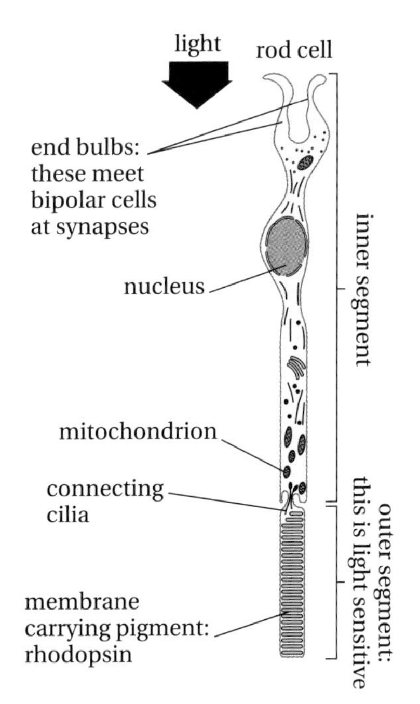

Fig 20
A rod cell

When rhodpsin is broken down, we say it is **bleached**. In bright light, most of the rhodopsin is bleached and we see mainly with our cones. When we go into a darkened room we see nothing for a few seconds because the cones do not work at low light intensities. However, after a while the rhodopsin is re-synthesised. As this happens, the eye becomes **dark adapted** and we can see using our rods, but only in black and white.

Visual acuity

Visual acuity is the ability to see in detail. For instance, these lines ═══ appear as two lines but of you look at them from a distance they appear to blend into one thicker line. The greater the distance at which we can distinguish two lines, the greater our visual acuity.

Although rods and cones both enable us to see, it is the cones that give us our high visual acuity. For example, if the letter E was to fall on the cones, we would clearly distinguish it as a letter E; if it were to fall on the rods, we would just see a blob (Fig 22).

The main reason for the difference in visual acuity between the rods and cones is **retinal convergence**. Fig 21 shows that several rods all converge on one neurone, but each cone has its own connection. Any impulses generated by an individual cone are transmitted to the brain but any impulses generated by a rod are fed into a neurone along with impulses from many others.

The greatest concentration of cones is at a point on the retina called the **fovea**. It is only when light falls on this point that we can see things in detail (Fig 22).

Fig 21
The concept of retinal convergence

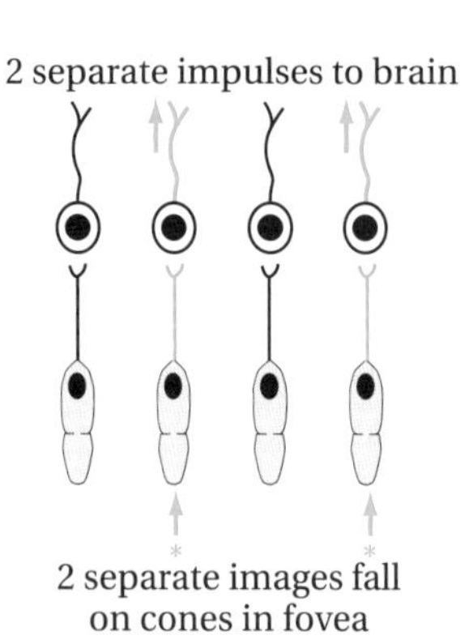

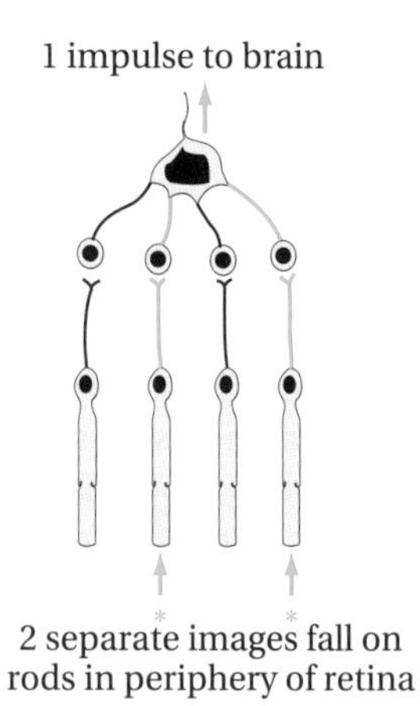

Fig 22
Visual acuity and the fovea

a When image falls on cones in fovea

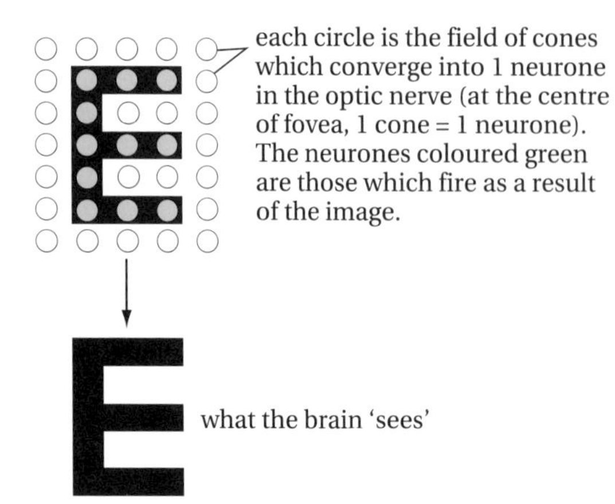

b When same image falls on rods in periphery of retina

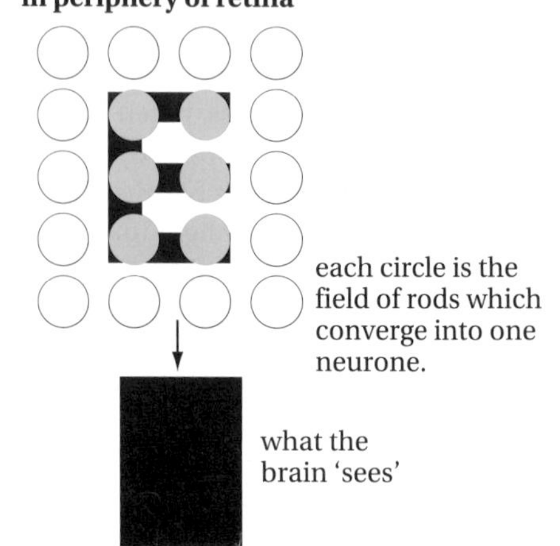

Colour vision

The cone cells are also responsible for our ability to see in colour. There are three types of cones, classified according to the type of iodopsin they contain. The three types of iodopsin are sensitive to different wavelengths/colours of light. Put simply, there are red-, blue- and green-sensitive cones.

Red light stimulates the red-sensitive cone but not the others, and so on. White light stimulates all three types, while black stimulates none of them. The different colours can be distinguished because they stimulate the three different types to a different extent. For example, yellow light stimulates both green- and red-sensitive cones.

It is difficult to tell the extent to which an animal can see in colour – they can't tell you what they can see. It is thought that primates evolved sophisticated colour vision due to their need to distinguish between a wide variety of foods, e.g. ripe and unripe fruit.

E

1 Candidates lose marks by saying that there are 'red cones' or 'green cones'. They are not coloured themselves, they are red-*sensitive*, green-*sensitive*, etc.

2 Many candidates state that the three types of cones are sensitive to red, blue and yellow, possibly due to their experience with mixing paint. With light, the primary colours are red, blue and *green*.

The nerve impulse

A **neurone** is a nerve cell, a specialised elongated cell that is capable of carrying impulses from one end to the other. There are several types of neurones but all have the same basic features:

- A **cell body** that contains the nucleus and other organelles.
- **Dendrites** that take impulses towards the cell body.
- An **axon** that takes impulses away from the cell body.
- **Synapses** – junctions with other neurones or effectors (e.g. muscles).

Fig 23 shows the basic structure of a neurone. Fig 24 shows how the neurone is involved in transmitting impulses.

E

Make sure that you can distinguish between a *neurone* (a single nerve cell) and a *nerve* (a bundle of axons surrounded by connective tissue).

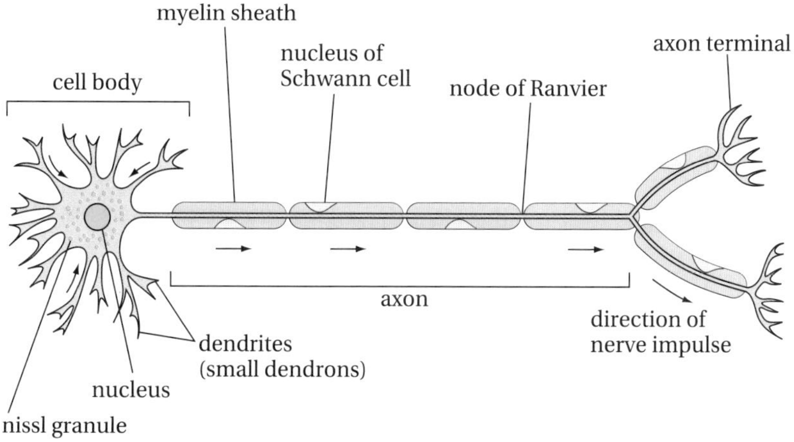

Fig 23
Basic structure of a motor neurone; all neurones have the same basic features

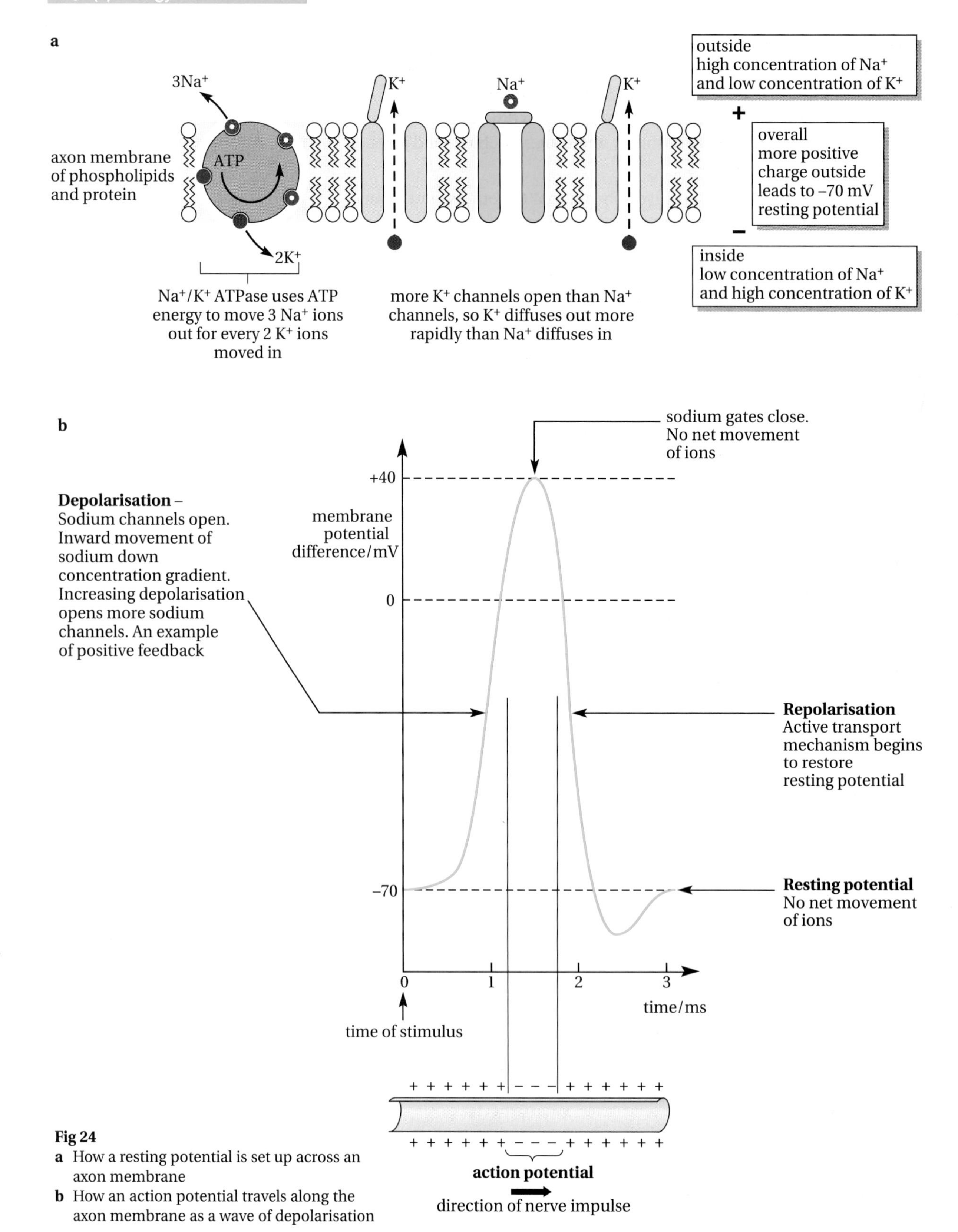

Fig 24

a How a resting potential is set up across an axon membrane

b How an action potential travels along the axon membrane as a wave of depolarisation

The resting potential

The resting potential is a charge across the axon membrane, caused by an unequal distribution of ions. This outside is positive with respect to the inside. Once the resting potential has been established, the neurone is ready to transmit an impulse. There are two steps in setting up the resting potential (see Fig 24a).

1. An active transport mechanism swaps positive ions; Na^+ ions go in, K^+ ions go out. In itself, this does not alter the charge across the membrane.
2. K^+ ions diffuse out more easily than Na^+ ions can diffuse in. This is because there are more open K^+ ion channels than Na^+ ion channels. This results in the accumulation of more positive ions outside the axon membrane, creating the resting potential.

The action potential

As Fig 24b shows, the action potential is a brief reversal of the resting potential, a change that travels rapidly along the axon. It is sometimes described as a wave of depolarisation.

The action potential can also be shown on an oscilloscope trace, see again Fig 24b. Exam questions often feature this diagram, so try to match the following steps to the diagram. The mechanism is as follows:

1. After the stimulus, sodium gates open and Na^+ ions flood inwards, reversing the polarity for a fraction of a second.
2. The sodium gates close, and the pump mechanism begins to re-establish the resting potential.
3. The sodium gates in the next region of the axon open, reversing the polarity in that region of the axon.

Always refer to action potentials as *impulses*, not messages. Messages are complex, whereas nerve impulses are simply electrical 'blips'.

E

Action potential facts:

- An action potential is not an electrical current, nor is it a 'message'.
- All action potentials are the same size, there are no large or small impulses.
- They travel along axons at speeds up to 120 metres per second.
- The speed of transmission depends on the axon diameter (larger = faster), the number of synapses in the pathway (more = slower) and whether the nerve is myelinated or not.
- Transmission is fastest in myelinated nerves, where the action potential jumps from node to node. This is **saltatory conduction**.

The brain interprets all incoming information by detecting:

1. The origin of incoming impulses. For example, it we hurt our left ankle, the brain knows what part of the body is affected because the impulses arrive down the nerve from the left ankle. Amputees sometimes have sensations in lost limbs because part of the nervous pathway is still intact
2. The frequency of the impulses – the more intense the stimulus, the greater the frequency.

Refractory period

This is the 'recovery period' after the transmission of an action potential. There are two phases:

1. The **absolute refractory period**. This is the time during which it is impossible to create another impulse, no matter how intense the stimulus.
2. The **relative refractory period**. This is the time during which it is possible to create another impulse, but the stimulus needs to be greater then normal.

The synapse

E Impulses do not cross synapses. An impulse comes to an end once it reaches a synapse. After synaptic transmission, a new impulse is generated.

Synapses are gaps between neurones. They are a vital component of the nervous system because they allow the selection of different pathways for transmission of impulses. All or our thoughts, memories, skills and actions are due to these elaborate pathways, and to their selection by the synapses.

The events of synaptic transmission are shown in Fig 25 and are described below:

1. The action potential arrives at the **synaptic knob**.
2. Calcium channels open, so Ca^{2+} ions flow into the synaptic knob.
3. This causes vesicles containing the transmitter to move to the **presynaptic membrane**.
4. The vesicles fuse with the presynaptic membrane and discharge their contents into the **synaptic cleft**.
5. Molecules of transmitter diffuse across the gap and fit into specific receptor sites on the **postsynaptic membrane**.
6. The permeability of the postsynaptic membrane changes, causing a movement of ions. Na^{+} ions flow inwards, building up a charge known as the **excitatory postsynaptic potential** (EPSP).
7. If the EPSP reaches a threshold, an action potential is generated in the neurone. Some transmitters cause **inhibitory** post synaptic potentials (IPSPs), which inhibit impulses. These are just as important as EPSPs in selecting particular neural pathways.
8. The transmitter substance it broken down by an enzyme in the cleft.
9. The products of breakdown are reabsorbed in to the synaptic knob, where they are re-synthesised using energy from ATP synthesised by the mitochondria.

All of the above events take about 0.01 second.

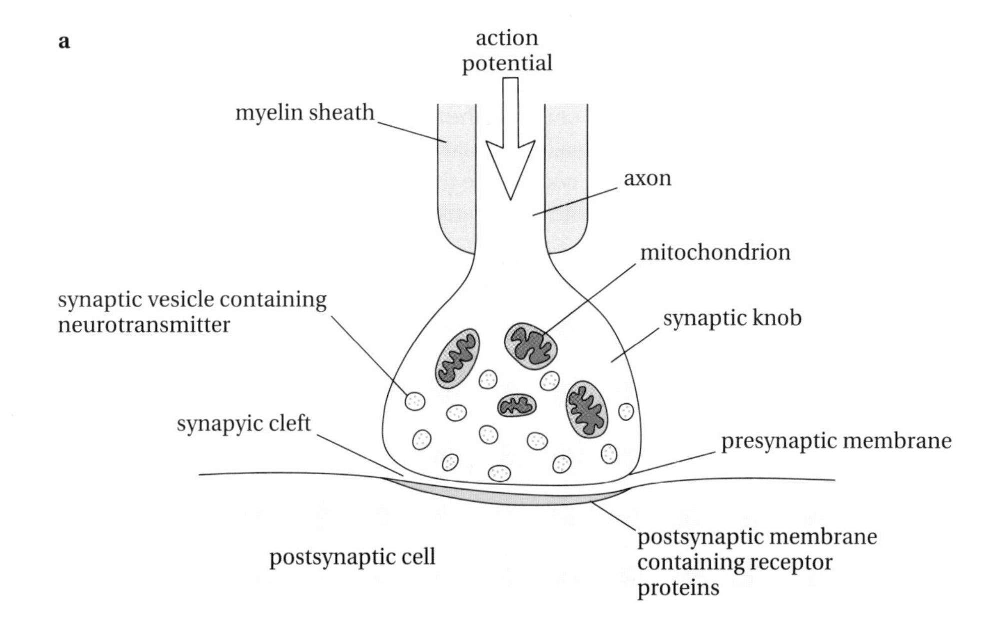

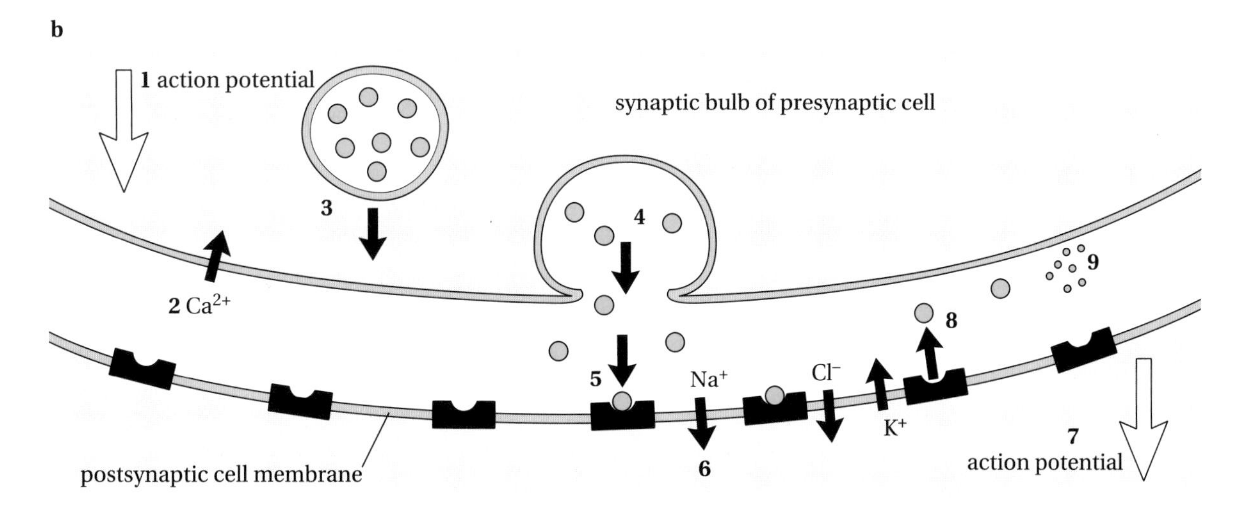

Fig 25
a The basic structure of the synapse
b The sequence of events in chemical transmission at a synapse

Many drugs and toxins work by affecting synaptic transmission in some way. You are not expected to learn the mode of action of a wide variety of drugs. Exam questions will test your understanding of synaptic transmission (outlined above) by using the example of a particular drug. As an example, try question 13 in the sample module test.

The most widespread transmitter is **acetylcholine**, which is found in the synapses of most voluntary nerves, including nerve–muscle junctions. The sympathetic nervous system uses **noradrenaline** as its transmitter, while the brain has many different ones, including **serotonin** and **dopamine**.

13.7 Analysis and integration

The brain and cerebral hemispheres

The brain is basically an enlarged front end of the spinal cord. It is a mass of neurones that coordinate the huge amount of sensory information that comes in from both the internal and external environment.

The cerebral hemispheres are the two large swollen areas that are the most obvious feature of the human brain. While the lower parts of the brain generally deal with the sub-conscious 'housekeeping' of the body such as the control of breathing and heartbeat, the cerebral hemispheres are concerned with the conscious activities.

Overall, the function of the cerebral hemispheres is to coordinate the incoming information, relate it to previous experience and select an appropriate response. Curiously, the left hemisphere controls the right side of the body and vice versa.

The sensory areas

Fig 26 shows that sensory information passes to the sensory area of the brain. Impulses from a particular part of the body stimulate a specific part of the sensory cortex.

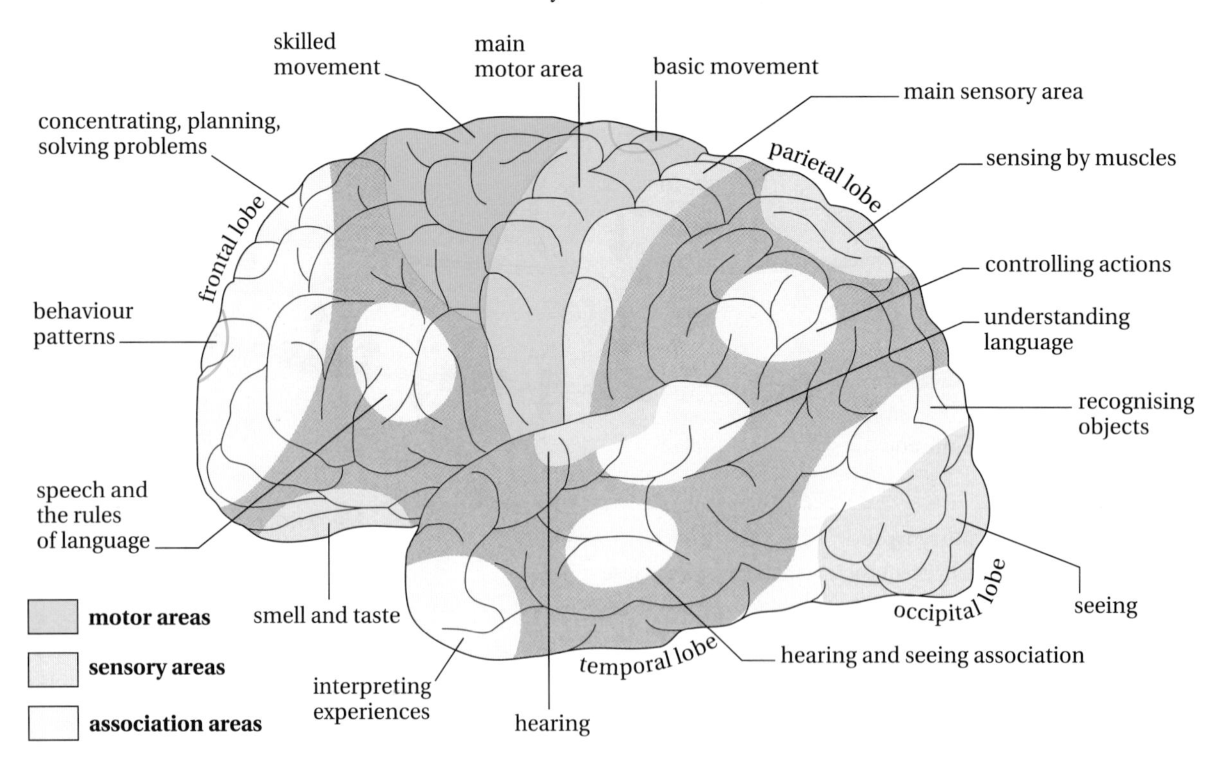

Fig 26
The sensory, motor and association areas in the human cerebrum. This diagram is mainly for reference; you will not normally be asked for this level of detail about the different areas.

The association areas

These areas allow an individual to interpret the sensory information in the light of previous experience, i.e. memory. The **visual association area** allows us to recognise things and put names to them. The **auditory association area** does the same for sound, allowing is to recognise a person's voice, for example.

The motor areas

As with sensory information, the motor area of the cortex is localised. Impulses to particular parts of the body pass from specific areas of the motor area. If a particular area of the brain is stimulated with an electrode, a specific muscle/ part of the body will move.

The autonomic nervous system (ANS)

The ANS is a branch of the nervous system that deals with many of the body's subconscious, or 'automatic' functions such as heartbeat, pupil size, etc.

The ANS consists of two sets of antagonistic nerves; the **sympathetic** system and the **parasympathetic** system. Generally, the impulses that travel down sympathetic nerves prepare the body for action, e.g. by increasing heartbeat, while impulses that travel down parasympathetic neurones bring the body back to normal e.g. by reducing heartbeat.

Table 3 shows the different effects of the two systems.

Function	Sympathetic effect	Parasympathetic effect
Heartbeat	Increases	Decreases
Pupil size	Dilates	Constricts
Salivation	Decreases	Increases
Bronchioles	Dilates	Constricts
Digestive juices	Inhibits	Stimulates
Tear ducts	No comparable effect	Increased secretion of tears
Bladder	Relaxes	Constricts

Table 3
The sympathetic and parasympathetic nervous systems compared

13.8 Muscles are effectors that enable movement to be carried out

Muscle is a specialised tissue that is capable of contracting when stimulated. Muscle tissue has the ability to contract with great force but it cannot expand again – it must be *pulled* back to its original length. For this reason, muscles usually work in **antagonistic pairs** or sets. In exams you may be given a situation where you must work out which muscles work antagonistically to each other – see question 11 on page 70.

There are three types of muscle in the mammalian body; **skeletal**, **smooth** and **cardiac**. An overview of the three types is given in Table 4. In this section, we look at the mechanism of contraction of skeletal muscle. You need to match the text to Fig 27.

Table 4
Summary of different muscle types

Function	Skeletal	Smooth	Cardiac
Other names	Striped, voluntary	Unstriped, involuntary,	Heart
Contraction	Rapid, powerful	Slow, less powerful	Rapid, powerful
Control	Voluntary	Involuntary	Involuntary. Myogenic; self generating
Speed of fatigue	Rapid	Slow	Slow (doesn't fatigue in a healthy person)
Distribution in the body	Attached to Skeleton	Mainly tubular organs, the gut, blood vessels, reproductive system.	Heart
Overall function in the body	Movement and maintenance of posture	Slow, often rhythmic squeezing known as **peristalsis**	To pump blood

Skeletal muscle

Fig 27d shows the basic structure of a **sarcomere** – one short length of a muscle fibre in which the pattern of bands is repeated. The changes in this banding pattern when muscle contracts reveals a lot about muscular contraction, making it a favourite topic for exam questions.

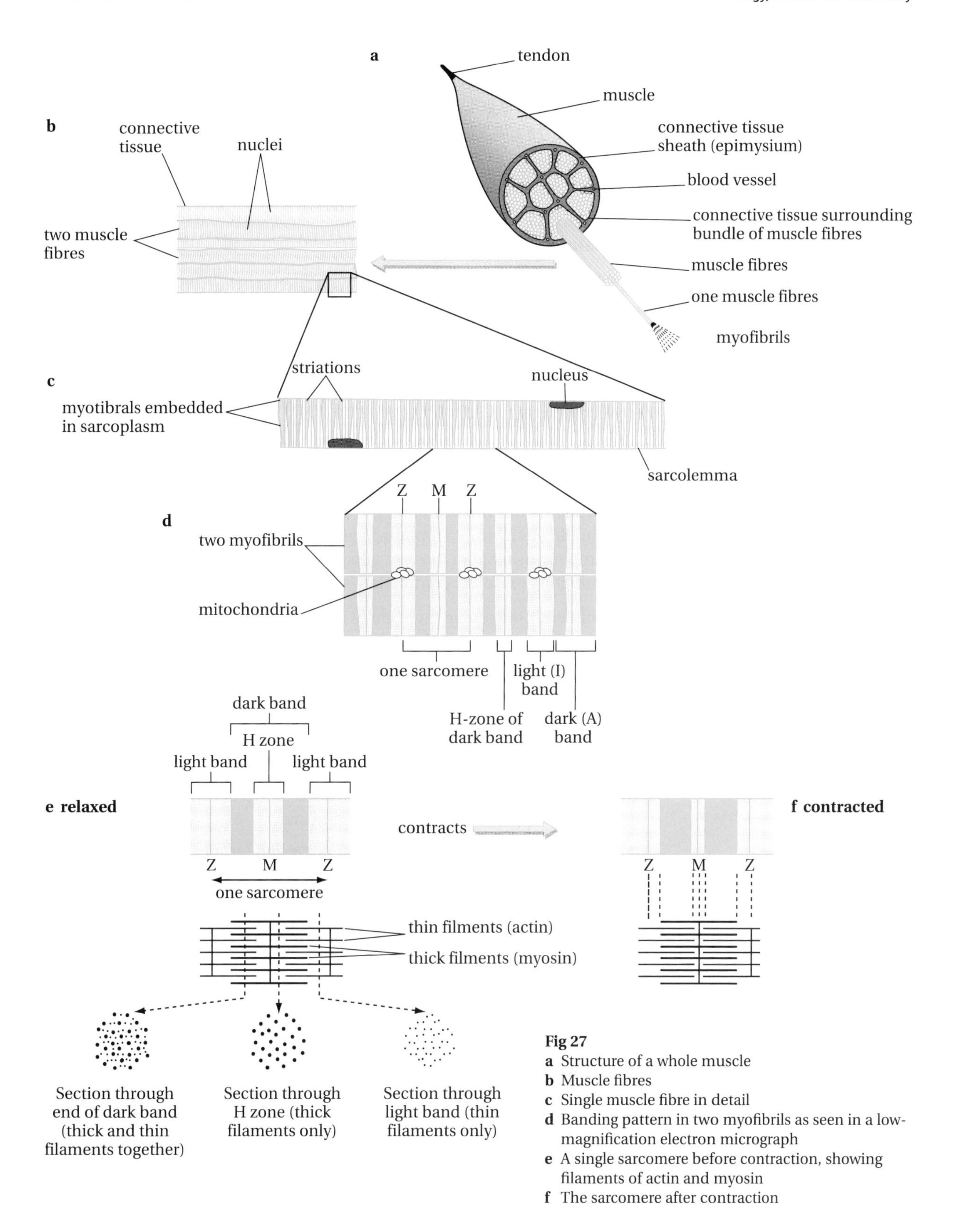

Fig 27
a Structure of a whole muscle
b Muscle fibres
c Single muscle fibre in detail
d Banding pattern in two myofibrils as seen in a low-magnification electron micrograph
e A single sarcomere before contraction, showing filaments of actin and myosin
f The sarcomere after contraction

Some skeletal muscle facts:

- An individual muscle is made up from hundreds of cylindrical **muscle fibres**, about 50 μm in diameter and ranging in length from a few millimetres to several centimetres.
- Muscle fibres are not composed of individual cells – the cell membranes have broken down and so each fibre has many nuclei.
- Each fibre is surrounded by a modified cell membrane called the **sarcolemma**.
- Each muscle fibre is composed many long cylindrical **myofibrils**, consisting of a repeating arrangement of proteins which causes the banding pattern.
- Each repeated pattern of proteins is called a **sarcomere**.
- The two main proteins involved in muscular contraction are **actin** (thin) and **myosin** (thick) filaments.
- Muscles contract when the actin fibres are pulled over the myosin fibres
- Two other proteins involved are **tropomyosin, and troponin**. Tropomyosin winds around the actin filament, preventing it from binding to myosin. Troponin, a globular protein, moves the tropomyosin out of the way, allowing actin to bind to myosin, thus initiating muscular contraction

Within each sarcomere there are two light bands; the I zone consisting of just actin filaments and the H zone that consists of just myosin. Between them are darker areas where these proteins overlap. When muscles contract, the actin and myosin filaments are pulled over each other so that the light bands get smaller. Note that the width of the dark band corresponds to the width of the myosin molecules, so it doesn't get any narrower - molecules don't shrink, they just slide over each other.

The sliding filament theory of muscle contraction

The basic steps in muscular contraction are:

1. An impulse arrives down a motor nerve and terminates at the **neuromuscular junction** (a modified synapse).
2. The synapse secretes acetylcholine.
3. Acetylcholine fits into receptor sites on the **motor end plate**.
4. The binding causes a change in the permeability of the **sarcoplasmic reticulum**, resulting in an influx of calcium ions into the **myofilament**.
5. The calcium ions bind to the troponin, changing its shape.
6. Troponin displaces the tropomyosin, so that the myosin heads can bind to the actin.
7. The myosin head pulls backwards, so that the actin is pulled over the myosin.
8. An ATP molecule becomes fixed to the myosin head, causing it to detach from the ~~myosin~~. actin

9 The splitting of ATP provides the energy to move the myosin head back to its original position, 'cocking the trigger' again.

10 The myosin head becomes reattached to the actin once again, but further along.

In this way, the actin is quickly pulled over the myosin in a ratchet motion, shortening the sarcomere, the whole filament and the muscle itself.

13.9 Inheritance

This section builds on the AS Module 2: **Genes and Genetic Engineering**. You need to be familiar with the structure of DNA and the basics of cell division (meiosis and mitosis).

Genotype

Some overall definitions:

D

- ***Genotype** is the genetic constitution of an organism. When we study the inheritance of one gene we use the term genotype to describe the combination of alleles the individual possesses; BB or Bb, for example.*
- ***Phenotype** is the observable features of an organism. The phenotype results from a combination of the genes it inherits and the environmental factors. For instance, we are all born with a certain combination of tallness genes so we have the potential to grow to a particular height. However, we will not reach this height without an adequate diet.*
- *Put simply; **Phenotype = genotype + environment***

You will also need to know the meaning of the following terms:

- **Gene** - a length of DNA that codes for the manufacture of a particular protein of polypeptide. In this section, it can be thought of as a piece of DNA that codes for a particular feature. Humans, like all other animals, are **diploid** organisms – we have two copies of each gene.
- **Chromosome** - one long DNA molecule with genes dotted along its length – see Fig 28.
- **Locus** - the position of a gene on a chromosome.
- **Allele** - an alternative form of a gene. Genes often have two or more alleles, e.g. flower colour gene could have two alleles, red and white.
- **Dominant** - the allele that, if present, is shown in the phenotype, e.g. the allele B could cause black fur in mice.
- **Recessive** - the allele that is only expressed in the absence of the dominant version, e.g. the allele b could code for brown fur in mice.
- **Homozygous** – when both alleles are the same. BB or bb.
- **Heterozygous** – when a pair of alleles is different; Bb.

Fig 28
Chromosomes contain many genes that are lengths of DNA that code for a particular protein

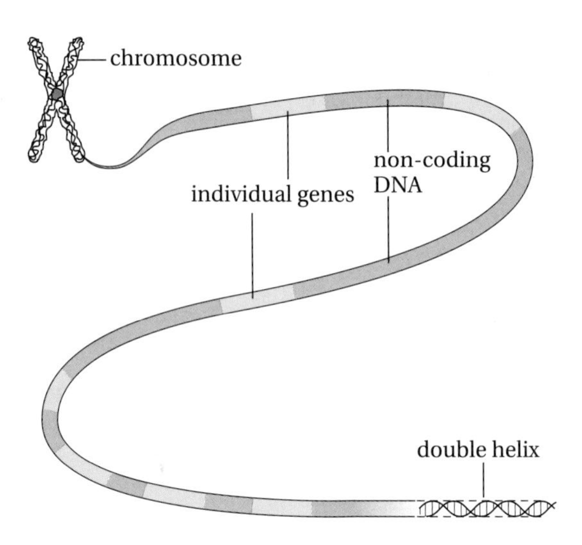

Meiosis and fertilisation

In AS Module 2 you learnt that meiosis is the special type of cell division that has two key features:

1 It halves the chromosome number.

2 It shuffles the genes so that all daughter cells are genetically unique.

In this section, we cover meiosis in more detail.

A key concept in understanding meiosis is that of the **homologous chromosome** (Fig 29). These are pairs of chromosomes – one originating from each parent. The two chromosome are matched in that they are the same length and in that they contain copies of the same genes. However, because they each come from a different parent, they are not absolutely identical. For example, at a certain locus (position) on one chromosome from the homologous pair there may be a coat colour gene that codes for brown fur. At the same position on the other chromosome may be a coat colour gene that codes for black fur. These two, slightly different forms of the same gene are called alleles.

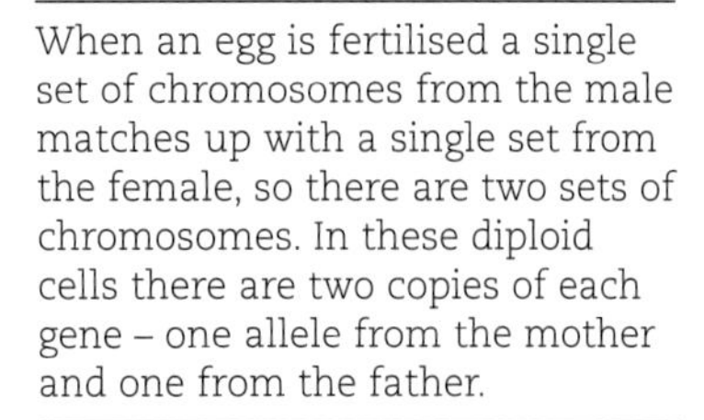
When an egg is fertilised a single set of chromosomes from the male matches up with a single set from the female, so there are two sets of chromosomes. In these diploid cells there are two copies of each gene – one allele from the mother and one from the father.

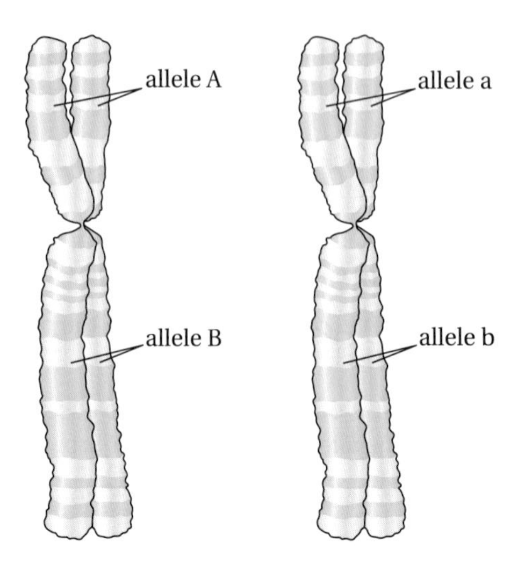

Fig 29
In diploid cells, chromosomes occur in pairs so that every cell has two alleles of each gene. In meiosis, only one chromosome from each homologous pair goes into each daughter cell. This shows how an individual can be genotype AaBb

Glossary of terms in meiosis

- **Bivalent** – a pair of homologous chromosomes lying alongside each other.
- **Chiasma** (plural = **chiasmata**) – points where homologous chromosomes attach during crossing over.
- **Chromatid** – one identical half of a double chromosome. During cell division, the chromatids are pulled apart to become single chromosomes.

The process of meiosis is described in Fig 30 and in Table 5.

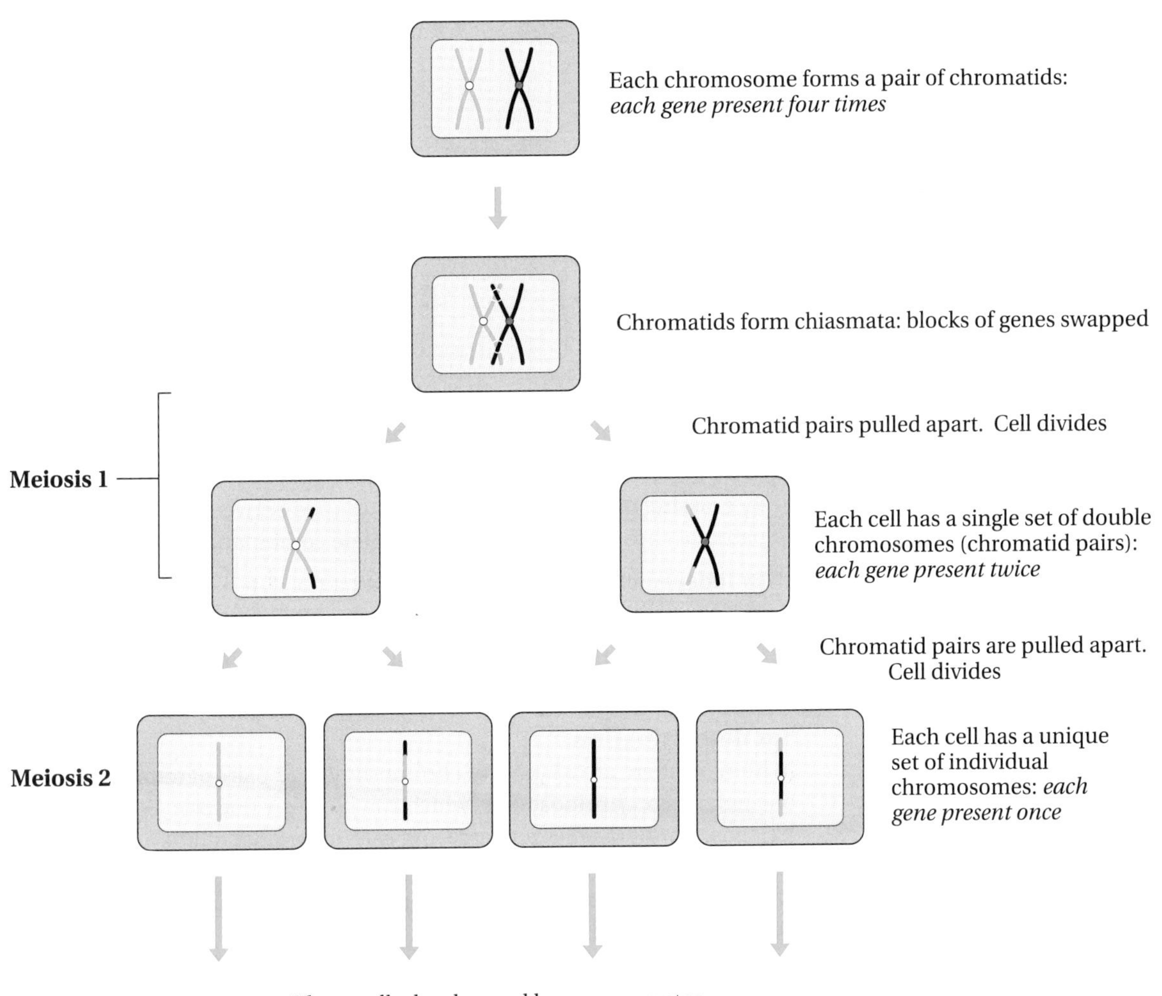

Fig 30
An overview of meiosis in an imaginary animal cell with just one pair of homologous chromosomes

Table 5
This table summarises the key events of meiosis that do not occur during mitosis. The first three events shown occur during the first division of meiosis. A second division then follows, forming 4 haploid cells.

Process	Diagram	Explanation
Pairing of homologous chromosomes		The homologous chromosomes lie alongside each other, forming a bivalent. This is a key event in meiosis and doesn't happen in mitosis
Chiasma formation		This is **crossing over** and it creates variation. Homologous chromosomes join at chiasma. When the chromosomes are pulled apart blocks of genes have been swapped
Separation of chromatid pairs from each homologous chromosome		The spindle fibres pull the two halves of the chromosome apart. Each new chromosome is a mixture of maternal and paternal alleles chromatids
Production of haploid cells		Haploid cells have one set of single chromosomes. These will develop into gametes. In animals, male gametes develop into spermatozoa and female gametes into eggs (ova)

Overall, sexual reproduction creates variation in three ways, the first two are concerned directly with meiosis:

1. By **crossing over**. Block is genes are swapped between maternal and paternal chromosomes, so that new gene combinations are created.
2. By **independent assortment**. One of a pair of chromosomes can pass into the gamete with any members of the other pairs.
3. By **random fertilisation** – for example any egg can combine with any sperm (in species that make eggs and sperm).

Sex determination

In humans there are 23 pairs of chromosomes; 22 pairs of **autosomes** and one pair of **sex chromosomes**, which can be X or Y (Fig 31). Females are XX; these chromosomes are large and carry a number of important alleles. Males are XY; the Y chromosome is much shorter with fewer genes, although it does have one or more vital genes that turn a developing embryo into a male.

The key to sex determination is meiosis in the male. The sex chromosomes, like the other pairs are separated by meiosis so that only one of each pair goes into the sperm. Thus, half the sperm contain an X chromosome, half a Y chromosome. All of the female's ova carry an X chromosome. If a Y sperm fertilises the egg, a male results; if an X sperm fertilises the egg, you get a female.

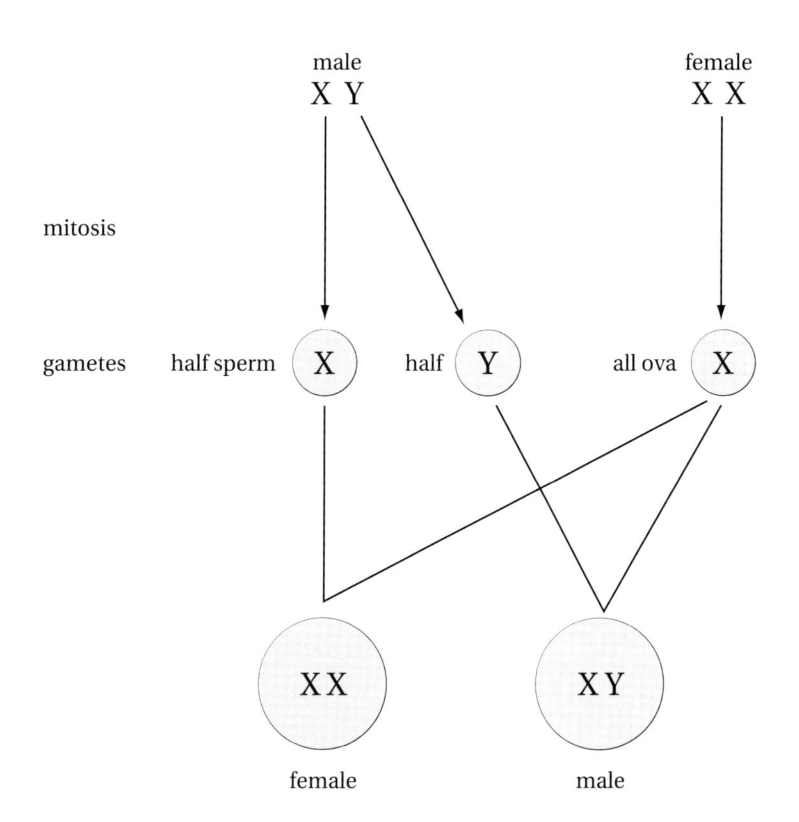

Fig 31
Sex determination in humans

Mendel's genetics – monohybrid and dihybrid inheritance

The basic rules of inheritance were worked out in the 1860s by Gregor Mendel. However, his work remained largely un-appreciated until around 1900, when the process of meiosis was first observed and understood. Meiosis went a long way to explaining Mendel's observations.

1 Monohybrid inheritance

This refers to the inheritance of a single gene with two alleles. For example, coat colour in a particular strain of mice is determined by one coat colour gene with two alleles, B (for black fur) and b (for brown fur).

There are three possible genotypes for these mice, but only two phenotypes:

Genotype	Phenotype
BB (homozygous dominant)	Black
Bb (heterozygous)	Black
bb (homozygous recessive)	Brown

Note that black is dominant because if the allele B is present, it is shown in the phenotype. The b allele is recessive, and is only expressed in the absence of the other allele. Fig 32 shows the basics of a monohybrid cross. If a Bb mouse mates with a Bb, there is a three-to-one ratio in the phenotypes of the offspring.

Fig 32
Monohybrid inheritance in mice. Pure breeding black and brown mice are homozygous (BB and bb). When bred, the mice born in the first litter are all black and are heterozygous (Bb). If these mice are then bred together, the litters have black and brown mice in a ratio of 3:1

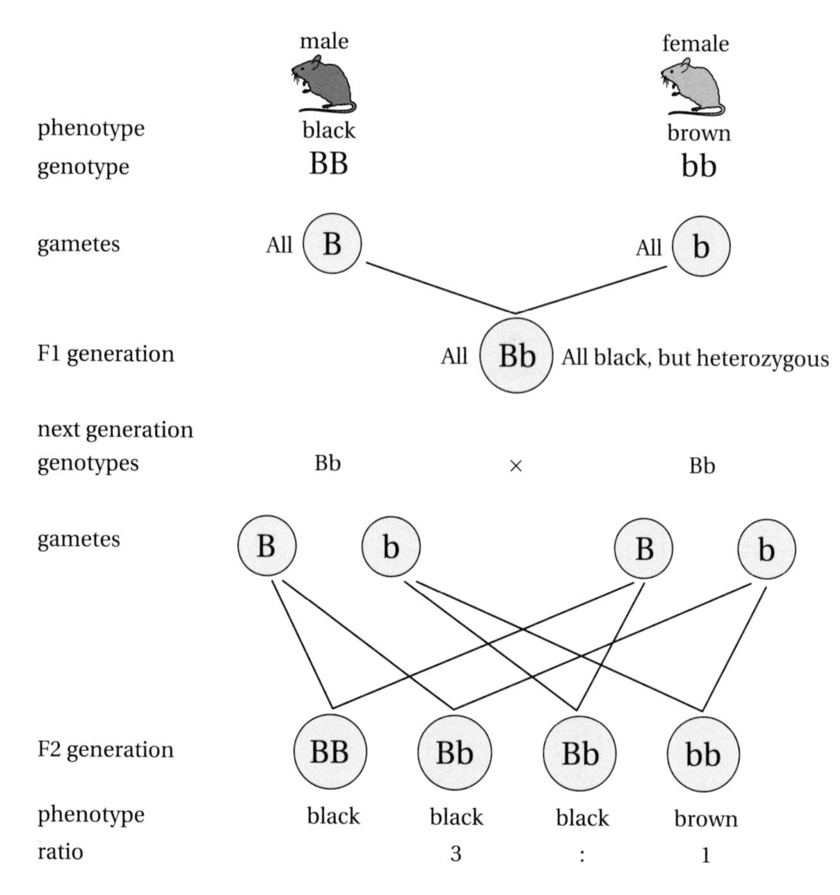

E

Every year candidates lose marks by writing ratios in the wrong way. A ratio of 1:3 is not the same as 1 in 3. 1:3 is the same as 1 in 4, and can also be expressed as .25 or 25%

In genetics we deal with large numbers of gametes, so the ratios are averages. It's not like the draw for the FA cup where you known what's been drawn and therefore what is left in the bag (as it were). If the chances of a child being born with genotype bb is 1 in 4, and the couple have three children of genotype BB, Bb and Bb, then the chance that the fourth will be bb is still 1 in 4.

Codominance

There are cases where alleles are **codominant**, where they are both be expressed in the phenotype. A classic example is the human ABO blood group system. This is controlled by one gene, I, with three alleles, I^A, I^B and I^O. I^A and I^B are codominant over I^O.

- Allele I^A codes for A proteins on the red cells.
- Allele I^B codes for B proteins on the red cells.
- Allele I^O codes for no relevant proteins on the red cells.

This gives us the table of genotypes and phenotypes shown in Table 6.

Table 6
The genotypes of blood groups

Genotype	Phenotype (i.e. blood group) and explanation
I^0, I^0	O – I^0 is recessive
I^A I^0 or I^A I^A	A – I^A is dominant over I^0
I^B I^0 or I^B I^B	B – I^B is dominant over I^0
I^A I^B	AB – I^A and I^B are codominant

Dihybrid crosses

These involve two separate genes, each with two alleles, at the same time. It is assumed that the genes are located on different chromosomes so that they can be separated by meiosis.

The vital steps in working out the products of a dihybrid cross are (see also Fig 33):

1. Parent's genotypes.
2. Gamete formation. This is where you can apply your knowledge of meiosis. One allele from each pair can go into the gamete, but by independent assortment it can pass into the gamete with any allele from the other pair. Thus a genotype of AaBb can give gametes of AA, Ab, aB and ab.
3. By random fertlisation any male gamete can combine with any female.
4. With dihybrid inheritance there can be up to 16 different genotypes, so it is important to organise the results using a Punnett square.

First cross

parents AABB × aabb

MEIOSIS

gametes all AB ab

F1 generation AaBb All black with short fur

When selfed

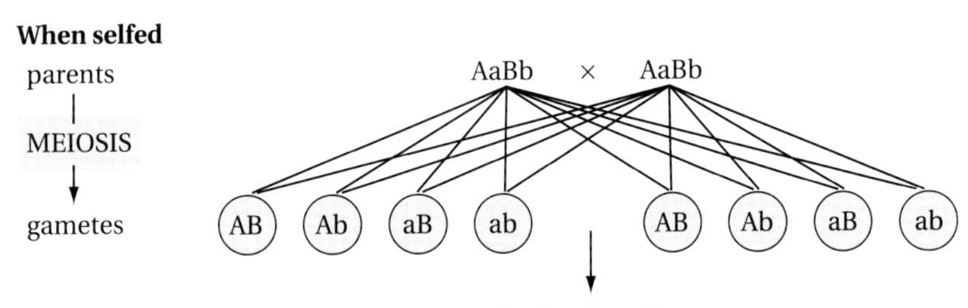

crosses organised using a Punnett square

female gametes

male gametes	AB	Ab	aB	ab
AB	AABB	AABb	AaBB	AaBb
Ab	AABb	AAbb	AaBb	Aabb
aB	AaBB	AaBb	aaBB	aaBb
ab	AaBb	Aabb	aaBb	aabb

F2 generation

Black, short fur	Black, long fur	brown, short fur	Brown, long fur
9	3	3	1

Fig 33
In mice, B, the allele for black hair is dominant to b, the allele for brown hair. A second allele, A, the allele for short hair, is dominant to a, the allele for long hair. A cross between two homozygous mice, on black with short hair and the other brown with long hair is an example of dihybrid inheritance. In the first generation, all the mice in the litter are black with short hair. In the second generation, black with short hair, black with long hair, brown with short hair and brown with long hair are produced in a ratio of 9:3:3:1

Epistasis

This is a situation where the presence of one allele affects the expression of another. A good example of this is seen in the coat colour of mice.

The natural grey/brown coat colour of wild mice is called **agouti**. The hairs are banded; dark at the base and lighter brown towards the tip. There are two genes involved, each with a dominant and a recessive allele:

- A/a is the gene for the ability to produce pigment. Allele A confers the ability to make pigment, while allele a does not.
- B/b is the gene for the ability to make banded hair. Allele B codes for the ability to make banded hairs, while allele b does not.

If the mouse is aa it will not be able to make pigment, so an albino mouse will result no matter what their genotype for the banding allele. However, if the mouse has allele A, the second B/b gene becomes relevant, as Table 7 shows.

Table 7
An explanation of the mouse hair types in this epistasis example

Genotype	Diagram	Phenotype	Explanation
AABB AABb AaBb AaBB		Agouti; mice with banded hairs	These mice possess both alleles A and B so can produce pigment *and* bands
AAbb Aabb		Black mice	Can make pigment but not bands
aaBB aaBb aabb		Albino mice	Do not possess allele A so can't make pigment. B/b becomes irrelevant

Sex linkage

Sex linked inheritance concerns genes on the sex chromosomes. The X chromosome is the larger of the sex chromosomes and contains more genes. With sex linked inheritance, the normal pattern is affected by the key fact that *males only have one X chromosome and so only have one copy of a sex linked allele.*

Haemophilia is a good example of a sex linked disease. This disease is cause by a faulty allele that fails to make the correct blood clotting protein, **factor 8**. There are two alleles; the normal H and the faulty allele, h. In sex linkage we show the chromosomes as well as the alleles.

There are three possible genotypes for females:

- X^HX^H = healthy; normal alleles on both X chromosomes.
- X^HX^h = healthy, but a carrier of the haemophilia allele.
- $X^h X^h$ = a haemophiliac (very rare in females).

Males have only two possible genotypes:

- X^HY^- = a healthy male. There is no allele on the Y chromosome.
- X^hY^- = a haemophiliac. He has no healthy gene on the Y chromosome to mask the effects of the faulty gene he has inherited.

A key point here is that, with sex linked inheritance, males can't be carriers.

Pedigree or 'Family Tree' diagrams

Pedigree means 'of known ancestry' and pedigree diagrams are often used in exams to test your understanding of inheritance and sex linked inheritance. Fig 34 shows the pedigree diagram in a recent exam question, together with the questions and model answers.

The diagram shows the inheritance of one type of myopia in three generation of a human family

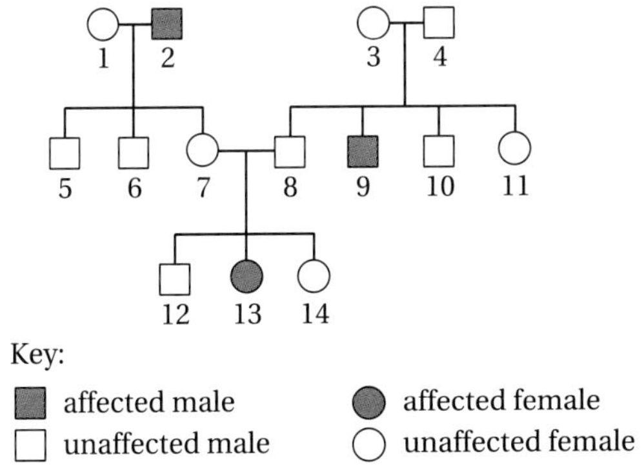

Key:
affected male
unaffected male
affected female
unaffected female

(a) Give **one** piece of evidence from the diagram which suggests:

(i) that themyopia is determined by a recessive allele;

Individuals 13 and 9 are affected, but not their parents who must be carriers

(ii) why it is **not** likely that myopia is a sex-linked condition.

Both males and females are affected

(b) Using A for the dominant allele and a for the recessive allele, give the possible geneotypes for:

(i) individual 7;

Aa

(ii) individual 12.

Aa or AA

Fig 34
Pedigree diagrams often feature in the exam – make sure you know how to interpret them

Sickle cell anaemia; a case study

Normal haemoglobin is coded for by the allele Hb^A. Sickle cells anaemia is a genetic disease in which a faulty allele, Hb^S makes an abnormal type of haemoglobin. In conditions of low oxygen tension, this type of haemoglobin polymerises into long chains, distorting the red cells into a sickle or banana shape. Sickle cells are not as flexible as normal red cells, and can cause blockages in small blood vessels, depriving areas of oxygen.

An interesting feature of the sickle cell allele is that it is much commoner than might be expected. This is because the heterozygotes have a survival advantage in areas where malaria is endemic.

There are three possible genotypes:

- Hb^A Hb^A = normal.
- Hb^A Hb^S = suffers from sickle cell, but more resistant to malaria.
- Hb^S Hb^S = a sufferer, likely to die early.

So while the homozygotes die, the heterozygotes have an advantage over the normal individuals, so the frequency of sickle cell disease remains high.

13.10 Variation

Types of variation

There are two basic types of variation; **discontinuous** and **continuous**.

Features that vary discontinuously fall into one category or another. The human ABO blood group system is a classic example. Individuals are either A, A AB or O – there is nothing in-between. Discontinuous features are usually controlled by single alleles, which the individual either has or it doesn't. Most of the cases of monohybrid and dihybrid inheritance that you study will feature discontinuous variation.

Features that vary continuously show a whole range of values. In humans most variation is continuous; height, shoe size, intelligence, etc. For most factors that vary continuously, most individuals fall into the mid range, giving a **normal distribution** (Fig 35). Continuous variation is usually controlled by several genes – it is said to be **polygenic**.

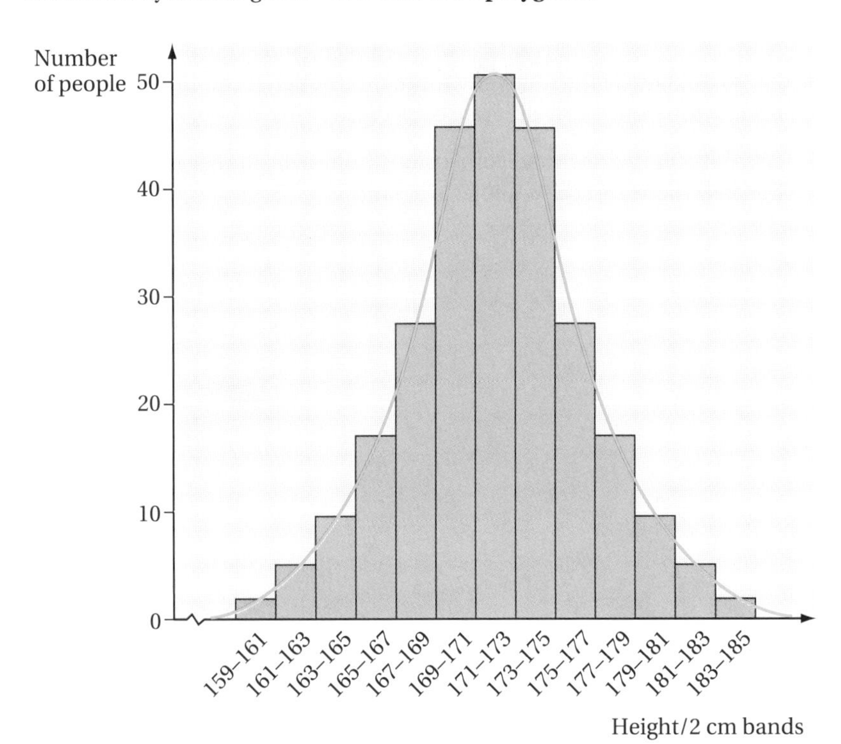

Fig 35
Height in human males is an example of variation. Most men's height falls in the middle, giving a bell-shaped curve known as a normal distribution

Causes of variation

As covered at AS level, the causes of genetic variation can be summed up as:

1. **Crossover** in meiosis.
2. **Independent assortment** in meiosis.
3. **Random fertilisation of gametes** – in animals, all sperm and ova are genetically unique, and which sperm fertilises the egg is totally random.

4. **Mutation**. Gene mutations are changes in the base sequence of the DNA so that the amino acid sequence changes and a different protein is made (see AS Module 2). This is usually harmful or irrelevant, but occasionally beneficial. Such beneficial mutations will give the individual a selective advantage – this is the driving force behind evolution.

Twin studies

When studying variation, it is difficult to determine the relative importance of genes and the environment. In humans, a lot of work has been done on identical and non identical twins, and this is a popular exam topic. The key points are:

1. Identical - or **monozygotic** – twins have exactly the same genotype so any differences between them must be due to environment i.e. their upbringing. Identical twins reared apart are especially interesting to geneticists.
2. Non-identical - or **dizygotic** - twins reared together have different genes but largely the same environment (though it can never be exactly the same). Any differences in these twins are largely genetic.
3. It is difficult to draw valid conclusions from twin studies in humans because there are so few subjects to study.

13.11 Selection and evolution

Natural selection

Variation is vital because it helps populations to survive unfavourable conditions, and in practice conditions are nearly always unfavourable. There is competition for resources; food, light, water, nutrients, etc. In addition, organisms have to survive drought, cold, heat, disease and a variety of other challenges.

D

- ***Natural selection is a process in which those organisms whose genes give them a selective advantage are more likely to survive, reproduce and pass their genes on to the next generation.***

Natural selection is often called *survival of the fittest,* but in a biological context, fitness means **reproductive success**. Fit organisms will pass more of their genes on to the next generation than less fit ones.

A good example of natural selection is that of antibiotic resistance in bacteria. What the media call 'superbugs' are bacteria that resist treatment with antibiotics. Some strains of bacteria are resistant to virtually all of the antibiotics at our disposal and they represent a growing threat.

E

Populations contain individuals with many different mutations that have accumulated over time. A change in the environment – such as a new antibiotic – will make a particular mutation relevant. Students often lose marks by stating that organisms mutate in *response* to the change.

Initially, the resistance probably arose as a result of a mutation. A single gene in a bacterium mutated so that it produced a protein that in some way made the bacterium resistant to the antibiotic.

It is important to appreciate that the antibiotic did not cause the mutation. However the mutation already existed in the population, or occurred in the right place at the right time, so it gave the individual a selective advantage. In areas of antibiotic use, the resistant bacteria multiplied, while the susceptible ones were killed. In this way, the frequency of the resistance allele increased over time.

Speciation

Defining a species is difficult because new species evolve out of existing ones and the process is gradual. However, for exam purposes a working definition of a species is:

D

- ***A population or group of populations of similar individuals that can mate and produce fertile offspring.***

Domestic dogs come in all shapes and sizes, but a Jack Russell and a poodle can mate to produce perfectly fertile crossbreeds. We say that the poodle and the Jack Russell are different breeds, not different species.

The classic example of this concept is the horse and the donkey. A male horse can mate with a female donkey to produce a mule. This hybrid, however, is sterile and so the horse and the donkey are regarded as separate species

The evolution of new species

There are three key steps to the evolution of a new species (see Fig 35):

1. **Isolation** Part of a population becomes isolated so that it cannot breed with the rest of the population.
2. **Natural selection** acts in different ways according to the local situation to change the frequency of alleles and, eventually, phenotypes.

3 **Speciation**. Over the generations, genetic differences accumulate so that the different populations can't interbreed, even if brought together again.

Fig 36
Steps in the evolution of a new species:

A population of rabbits on an island is split by a new river, so the two populations are isolated.

Natural selection acts in different ways on the two populations. In the rocky area, the rabbits can't burrow, so are more vulnerable to predators. Natural selection favours those with the keenest senses, best reflexes and greatest speed. They evolve into hare-like creatures.

Over the generations, genetic differences accumulate so that even if the two populations come into contact, they are not able to breed successfully.

13.12 Classification

We know of the existence of over two million known species of organism, but a large number are yet to be classified. The science of classification is known as **taxonomy**, and it aims to produce a catalogue of living things on Earth.

All species are given a **scientific name**, which usually comes from Latin or Greek. This name is used across all language barriers, and avoids confusion when referring to a particular species.

Scientific names are **binomial** – they have two parts. The first part, the **generic name**, is the name of the **genus** and has a capital letter. The **specific name** follows, and does not have a capital letter e.g. *Canis familiaris* – the domestic dog, or *Gorilla gorilla* – the gorilla. Scientific names should be given in italics or underlined.

This system of classification is called **phylogenetic** – i.e., based on evolutionary history. Its a bit like a family tree that goes back millions of years. To construct a phylogenetic tree, scientists use anatomical/physical features, fossil records and biochemical analysis of DNA and proteins.

The taxonomic hierarchy is shown in Table 8. Its key features are given below:

- It consists of a series of groups within groups, from the most general (kingdom) to the most specific (species).
- There is no overlap between the groups. For instance, there is no organism that is part amphibian and part reptile – its either in one group of the other.
- The groups are based on shared features. The more specific the group, the more shared features there are.

Table 8
Taxonomic hierarchy

Taxon	Humans as an example	Explanation
Kingdom	Animalia	We are animals – see the five kingdoms explanation below
Phylum	Chordates	We possess a backbone
Class	Mammals	We have warm blood, fur and feed our young on milk
Order	Primates	We have large brains, grasping hands fingernails binocular colour vision
Family	Hominids	Manlike creatures, only one species not extinct
Genus	*Homo*	Literally 'man'
Species	*sapiens*	Literally 'thinking'

It may help to invent a mnemonic to remember the sequence of kingdom, phylum etc. make up a phrase according to KPCOFGS, such as Keep Putting Cabbages On Fat Greasy Slugs, or something more memorable/personal to you.

The five kingdoms

Scientists have recently accepted the five kingdom system of classification. All organisms can be placed into one of these five kingdoms. Four contain the **eukaryotic** organisms while a fifth contains the **prokaryotes** – the bacteria.

The five kingdoms and their essential features are:

1 Kingdom Animalia (animals)

- Includes all the vertebrates (mammals, birds, reptiles, amphibians, fish) and invertebrates (insects and other arthropods, molluscs, segmented worms, roundworms, tapeworms/flukes and jellyfish).
- Animals are eukaryotic and multicellular. Their cells have no walls.
- They cannot make food, so move in search of it. They have muscles and nervous systems.

2 Kingdom Plantae (plants)

- Includes flowering plants, conifers and ferns along with simpler plants such as mosses, liverworts. NB Not the algae/seaweeds.
- Eukaryotic and multicellular. Cells have walls made from cellulose. Usually a large vacuole in mature cells.
- Most plants have leaves, stems and roots.
- Most plants possess chlorophyll and make food by photosynthesis. A few are modified to a parasitic way of life.

3 Kingdom Fungi

- Includes filamentous fungi and yeasts.
- Single celled fungi are called yeasts.
- Eukaryotic, but the multicellular fungi do not have separate cells – the nuclei are dotted around in the tissue.
- Walls of fungal cells contain chitin.
- All fungi feed by extracellular digestion – secreting enzymes and absorbing the soluble products. Most feed on dead material, i.e. are saprophytes. Some are parasitic.

4 Kingdom Protoctista

- The members of this kingdom seem to have very little in common, more like an assembly of organisms that could not be put into any other kingdom – a 'taxonomic dustbin'. The scientific reason for this group is based on their ancestry and evolutionary relationships.
- Includes single cells protozoans such as amoeba and plasmodium (the malaria parasite, the algae (including seaweeds) and sponges.
- Their cells are eukaryotic, but show a diversity of cell types.
- They feed in a variety of ways.

5 Kingdom Prokaryotae

- Includes the bacteria and single celled, prokaryotic organisms.
- The cells of prokaryotic organisms are much smaller than those of eukaryotes.
- Most feed by extracellular digestion – secreting enzymes and absorbing the soluble products. Most feed on dead material, i.e. are **saprophytes**. Some are **parasitic**. The cyanobacteria are photosynthetic.
- Prokaryotes reproduce mainly by binary fission (splitting in half), but can reproduce sexually.

You should remember the differences between prokaryotic and eukaryotic cells from AS Module 1. Eukaryotic cells are large with a lot of internal organisation i.e. many organelles made from membranes. Prokaryotic cells are much smaller – about a thousandth of the volume, and have little internal organisation.

E

Don't confuse protoctistans with prokaryotes. The words look similar but protoctistans are a kingdom of eukaryotes

NOTES

A2 4 Sample module test

Section A

1 The diagram below shows part of the respiratory pathway. The broken arrows represent intermediate stages.

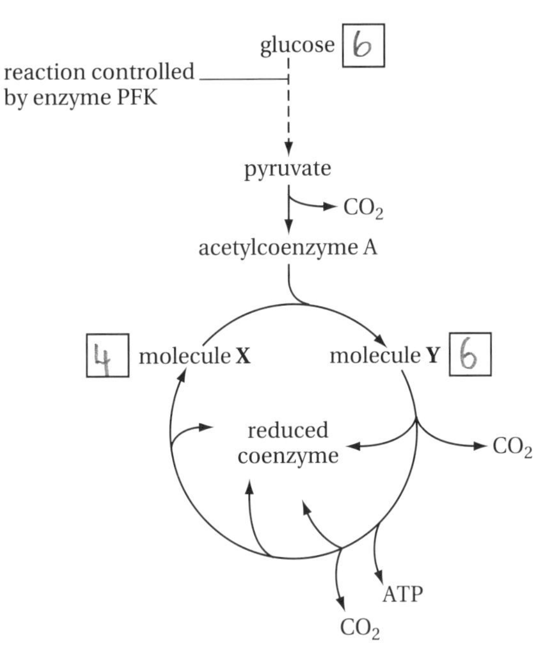

a Write in **each** of the three boxes on the diagram the number of carbon atoms in the molecules indicated. *(1)*

b Give **two** different chemical processes which are involved in the conversion of Molecule **Y** to Molecule **X** in the Krebs cycle.

1 reduction/oxidation

2 dephosphorylation *(2)*

c ATP regulates its own production in respiration by affecting the activity of the enzyme PFK. Suggest how this regulation may be achieved through negative feedback.

ATP levels drop sensed by receptor, which sends signals to mitochondria, which produces more ATP. When more ATP produced receptor no longer sends signals to mitochondria

(2)

2 The diagram below summarises the biochemical pathways involved in photosynthesis.

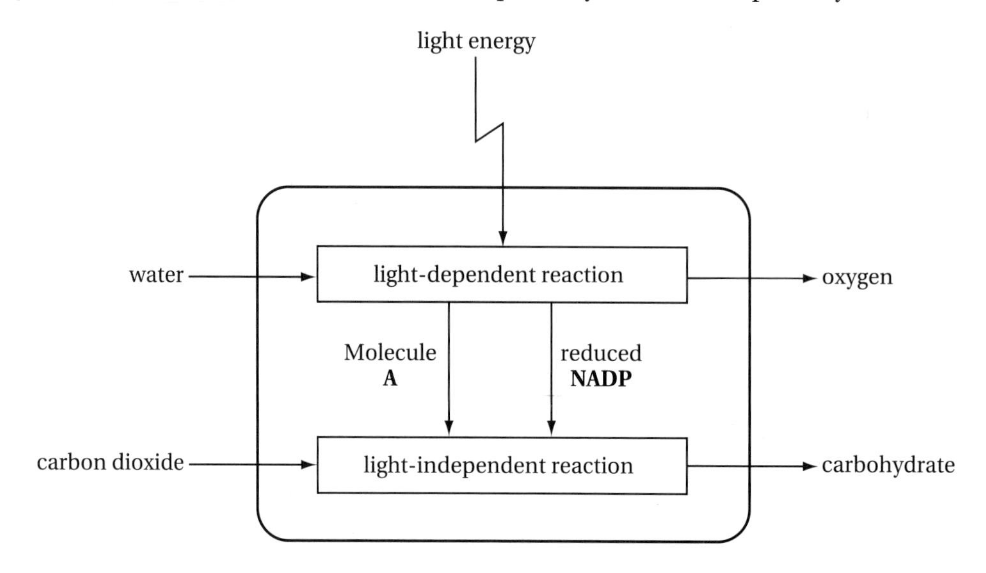

a Name Molecule **A** ..glucose.. *(1)*

b i Describe how NADP is reduced in the light dependent reaction.

....

....

....

.... *(2)*

ii Describe the part played by reduced NADP in the light-dependent reaction.

....

....

....

.... *(2)*

3 The histogram below shows the height of wheat plants in an experimental plot.

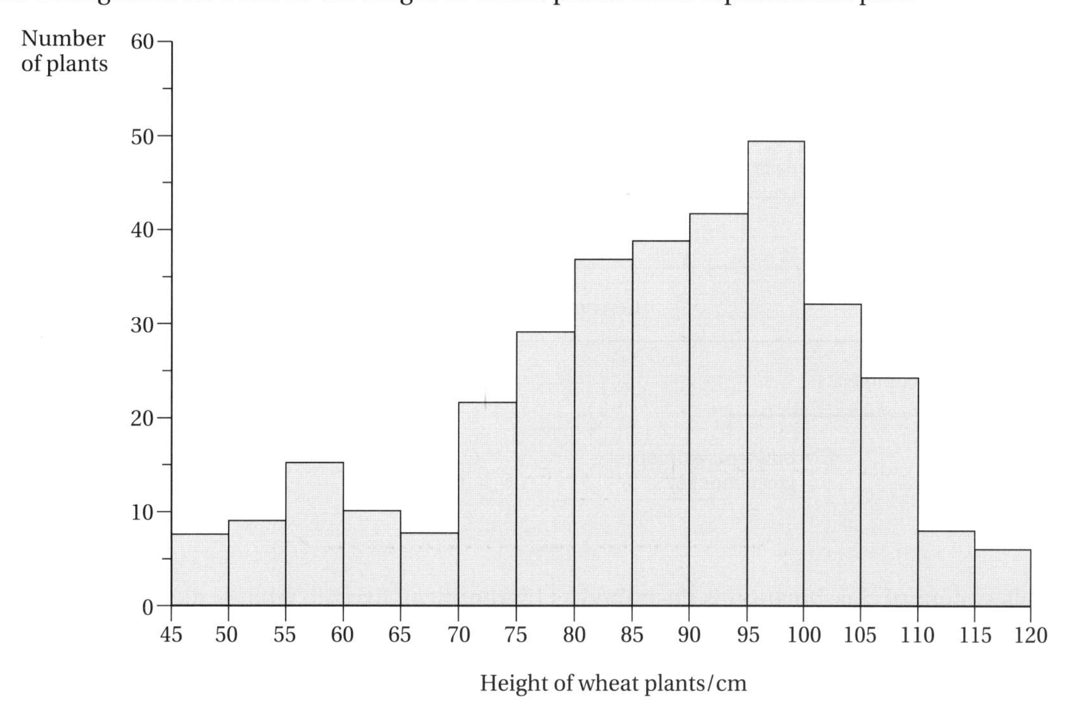

a What evidence from the data suggests that there are two strains of wheat growing in the experimental plot?

Two different peaks.

(1)

b i Which type of variation is shown by the height of each of the strains of wheat plants?

continuous.

Give the reason for your answer.

variation in height not a single height

(2)

ii Explain why the height of the wheat plants varies between 45 cm and 120 cm.

environmental differences.

(1)

4 The diagram below shows the way in which four species of monkey are classified.

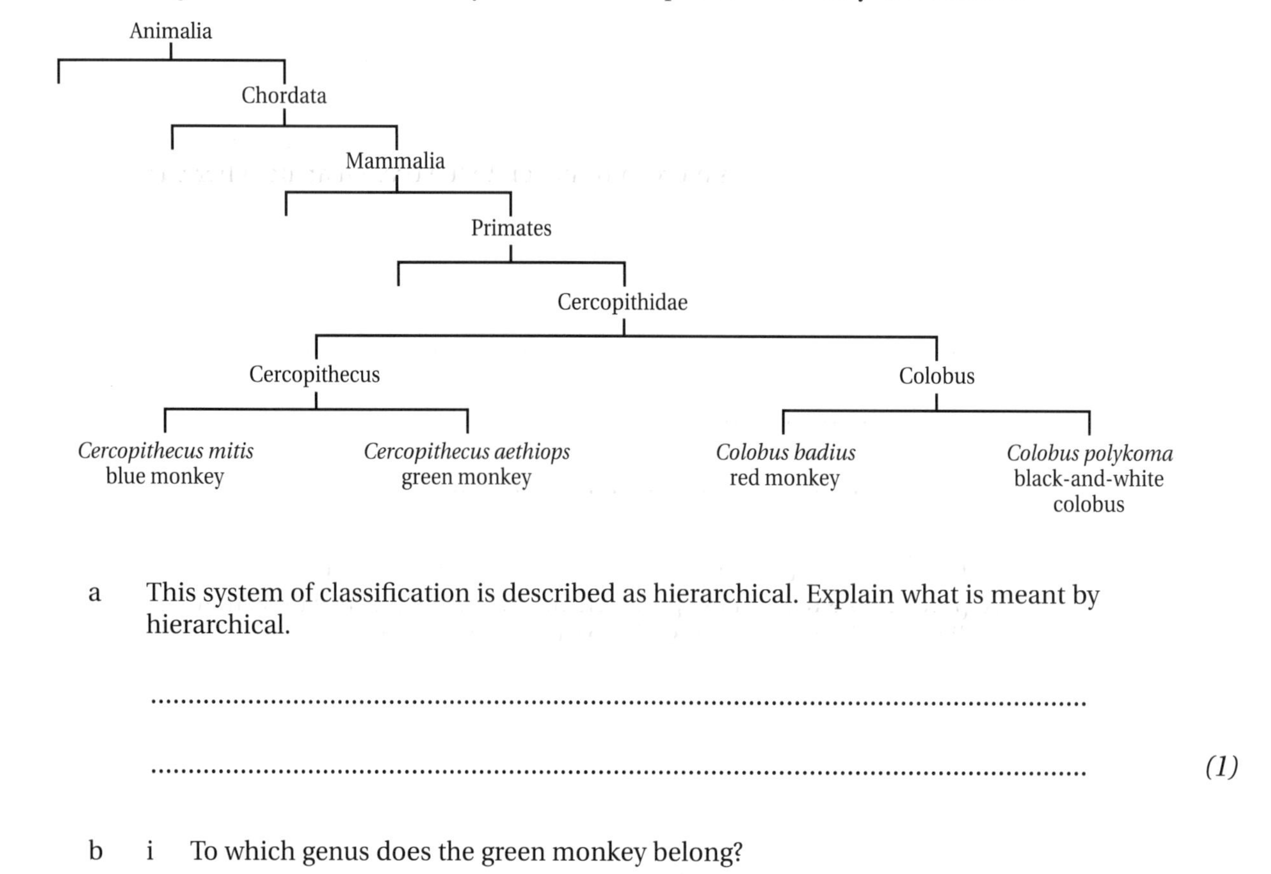

a This system of classification is described as hierarchical. Explain what is meant by hierarchical.

....

.... *(1)*

b i To which genus does the green monkey belong?

.... *(1)*

ii To which family does the red colobus belong?

.... *(1)*

c What does the information in the diagram suggest about the similarities and differences in the genes of these four species of monkey?

....

....

....

.... *(2)*

5 a Explain how the eye is able to detect an object that is purple.

.... *(1)*

b When we look directly at an object, its image is focused on part of the retina known as the fovea. Explain how the fovea provides more information to the brain than do other parts of the retina.

....

....

....

.... (2)

c The eyes of short-sighted people focus light rays from a distant object in front of the retina instead of on it, producing a blurred image on the retina. Laser surgery can be used to correct this defect. The diagram shows the results of one treatment in which part of the cornea is removed by laser surgery. After the surgery the surface of the cornea heals to give a new, smooth surface, with a smaller curvature.

Transverse section of cornea

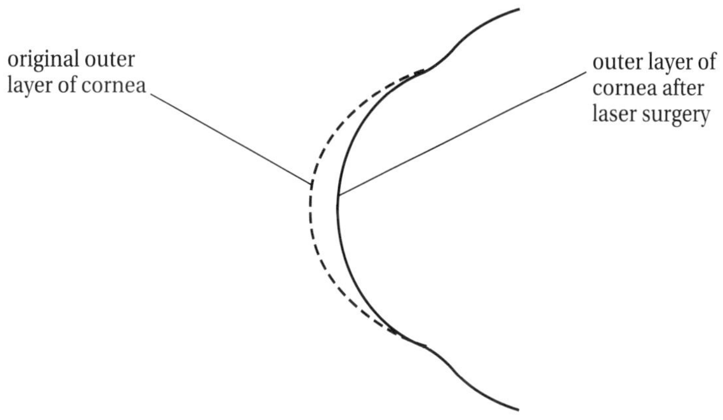

Explain how the surgery corrects the patient's short-sightedness.

....

....

....

.... *(2)*

d Blindness can be caused by a cataract, a condition in which the lens of the eye becomes cloudy. If the lens is removed, the patient can see again, but vision is not as good as before the cataract developed. Explain why this is so.

The light can no longer be focused onto the retina. Therefore a clear picture can not be generated

(2)

6 a Describe the role played by calcium ions in muscle contraction.

calcium ions flood into myofibrils binding to troponin, which changes shape moving the tropomyosin away from the actin active sites, enabling the myosin heads to attach. (2)

b The graph shows the force generated by a muscle plotted against the mean sarcomere length. The diagrams show the appearance of the sarcomere at points A and B.

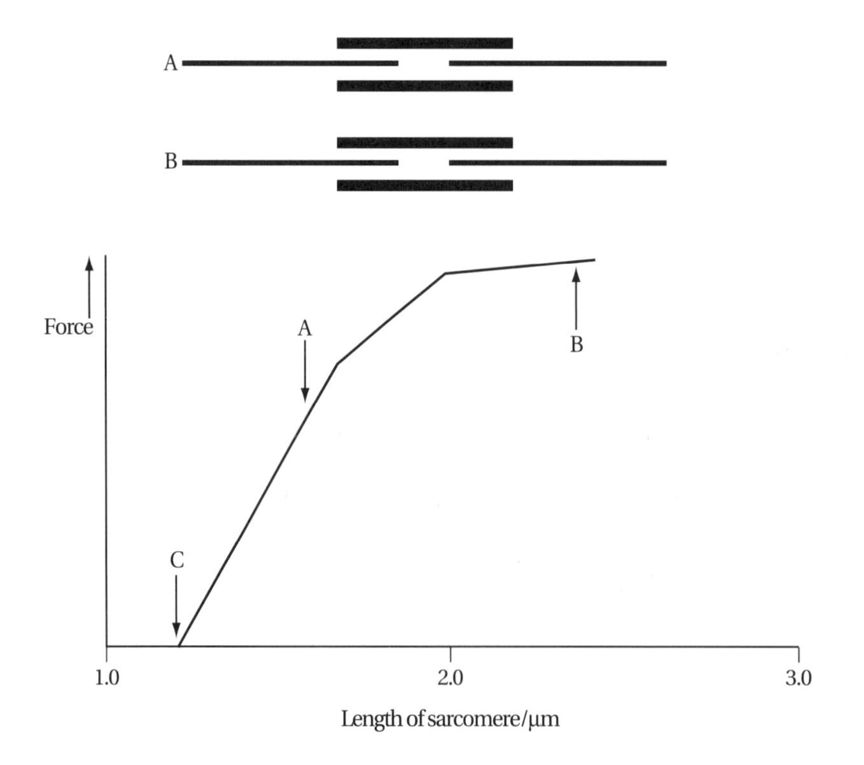

i What is the relationship between the force generated by the muscle and the number of actomyosin bridges? Explain your answer.

more bridges result in more force

Shorten length of sarcomere (2)

ii Draw a simple diagram to show the appearance of the sarcomere at point C on the graph. (1)

7 What is the function of **each** of the following parts of the cerebral hemispheres?

i Sensory areas

to create an image /sound from recieved impulses

ii Association areas

recognise images or sounds, by remembering previous events

iii Motor areas

send impulses to certain muscles resulting in muscle movement (3)

b In 1861, a man called Leborgne, suffered a stroke. As a result of this he was unable to speak and was paralysed on his right side. The drawing shows the location of the damage found in Leborgne's brain.

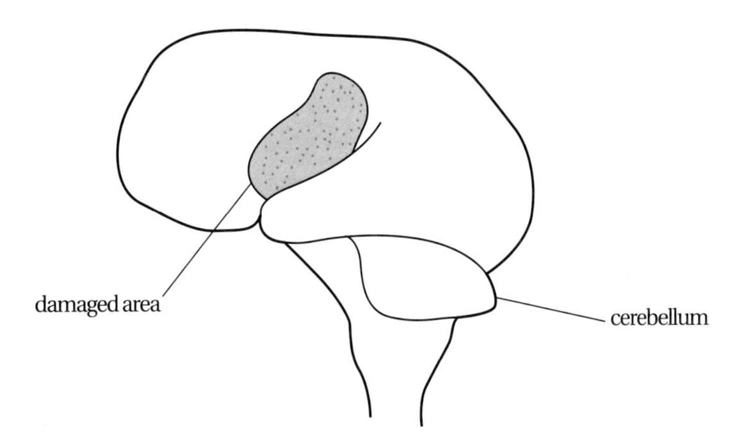

Suggest an explanation for:

i Leborgne being unable to speak.

damage to association area so can't decide what words to use

ii Leborgne being paralysed on his right side.

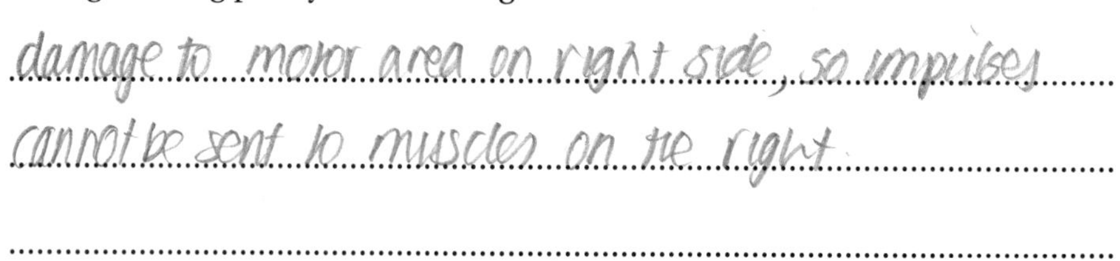

(2)

8 The inheritance of colour in sweet peas is controlled by two pairs of alleles (**A** and **a**, and **B** and **b**) at two gene loci on different chromosomes. The diagram below shows how the dominant alleles at these two gene loci determine flower colour by controlling the synthesis of a purple pigment.

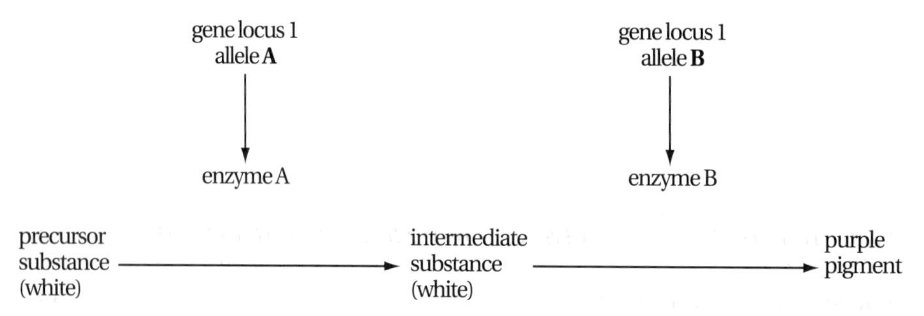

a Use information in the diagram to explain why plants homozygous for the mutant allele, **a**, have white flowers.

The ~~recess~~ dominant alleles determine colour. The recessive allele is unable to synthesise Enzyme A. resulting in the precursor substance not been converted to intermediate substance. ~~Bro~~ Stopping the enzyme pathway

(3)

b Use a genetic diagram to explain the genotypes and the expected phenotypes when plants with the genotypes **Aabb** and **aaBb** are crossed.

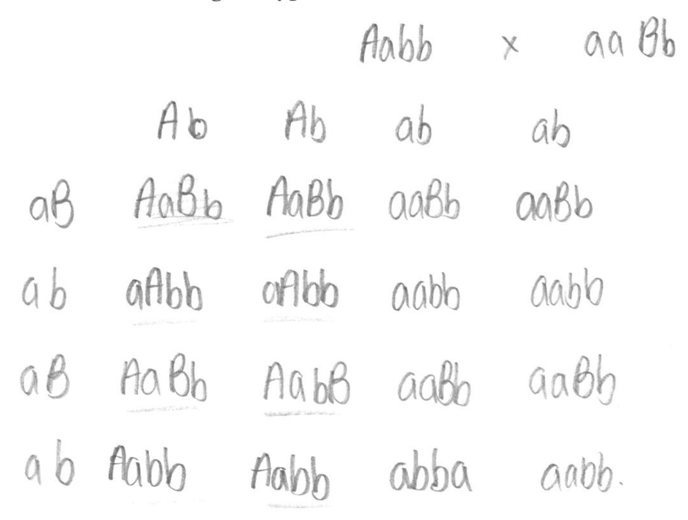

4 x purple.
4x intermediate white
8x precursor white

(4)

9 Rats and mice are common pests. Warfarin was developed as a poison to control rats and was very effective when first used in 1950. Resistance to warfarin was first reported in British rats in 1958 and is now extremely common. Warfarin resistance in rats is determined by a single gene with two alleles $\mathbf{W^s}$ ad $\mathbf{W^r}$. Rats with the genotypes listed below have the characteristics shown.

$\mathbf{W^sW^s}$ Normal rats susceptible to warfarin.

$\mathbf{W^sW^r}$ Rats resistance to warfarin needing slightly more vitamin K than usual for full health.

$\mathbf{W^rW^r}$ Rats resistant to warfarin but requiring very large amounts of vitamin K. They rarely survive.

a Explain why:

i there was a very high frequency of $\mathbf{W^s}$ alleles in the British population of wild rats before 1950;

(1)

ii the frequency of $\mathbf{W^r}$ alleles in the wild rat population rose rapidly from 1958.

the rats with W^r survived whereas W^s rats did not. W^r rats bread successfully

W^s rats disadvantage.

(2)

b Explain what would be likely to happen to the frequency of $\mathbf{W^r}$ alleles if warfarin were no longer used.

decrease as W^S rats survived easily as they do not require increased vit K.

(2)

c In humans, deaths from conditions where blood clots form inside blood vessels occur frequently in adults over the age of fifty. These conditions may be treated successfully with warfarin. However, some people possess a dominant allele that gives resistance to warfarin. Why would you not expect this allele to change in frequency?

past breeding age.
death past 50

(2)

10 The coyote, jackal and dingo are closely related species of the dog family. Their distribution is shown on the map.

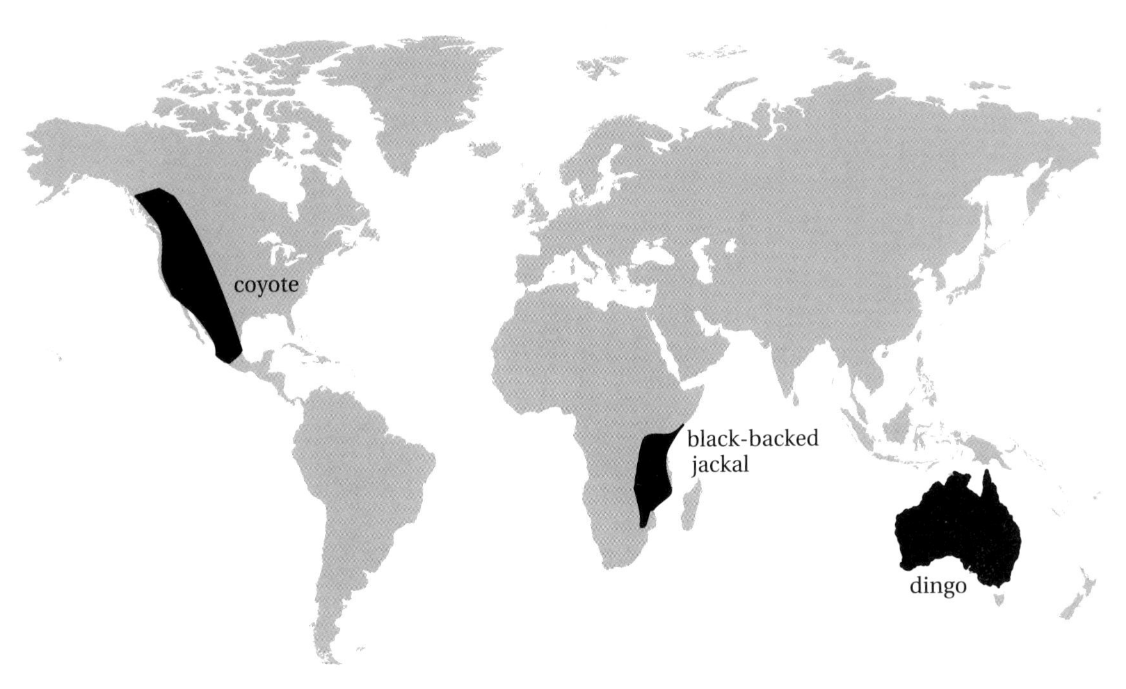

Suggest and explain how these three distinct species evolved from a common ancestor.

....

....

....

....

....

....

....

....

....

.... *(4)*

11 The diagram shows the main muscles in a human leg.

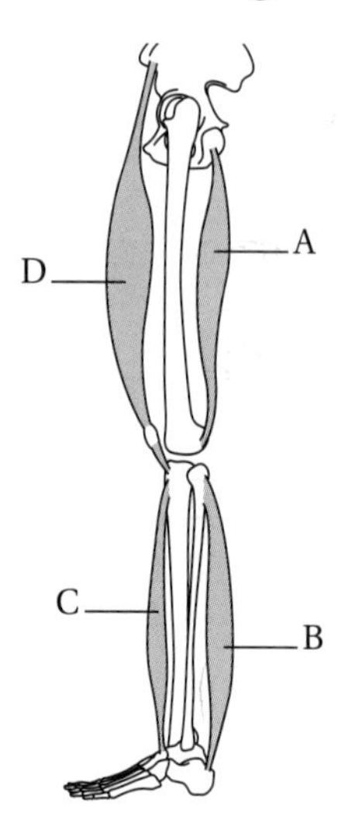

a Which of muscles **A** to **D** on the diagram

i must contract to raise the heel from the ground;

B

ii is antagonistic to this muscle?

C

2

b While walking barefoot on the beach, a person stands on a sharp stone.

i Briefly describe the pathway of the reflex arc involved in responding to this stimulus. (You are not required to give details of the propogation of nerve impulses.)

sensory neurone from recepter

~~motor~~ relay neurone in spinal cord

motor neurone to muscle.

3

(ii) Suggest one advantage of reflexes such as this one.

restrict damage

1

12 The diagram shows an organelle from a palisade mesophyll cell, as seen with an electron microscope.

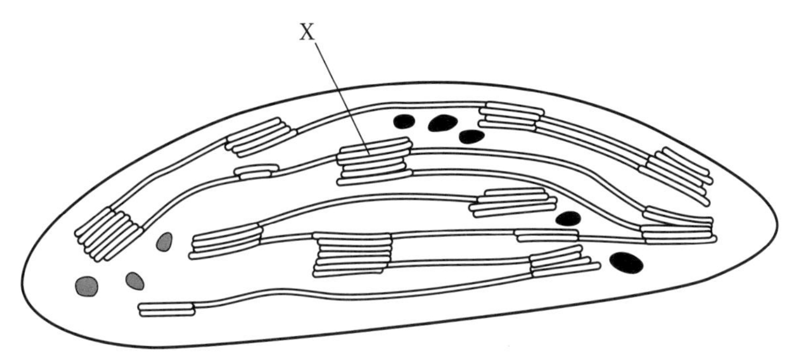

Magnification: × 38 000

a Name

i the organelle;

chloroplast

ii the part labelled **X**.

thylakoid

2

b Calculate the maximum length of the organelle in micrometers. Show your working.

Length.............micrometres (μm)

2

c Give **two** ways in which the structure of this organelle is adapted for its function.

1 ....

....

2 ....

....

2

Section B

13 Read the following passage.

Arrow poisons were widely used for hunting in South America where most of the powerful poisons such as curare originated. Charles Waterton carried out experiments on curare in the 1920s. He gave curare to a donkey that appeared to die ten minutes later. The animal was then revived by artificial ventilation of its lungs with a pair of bellows and went on to make a full recovery. The experiment showed that injection of arrow poison into the blood stream causes death by respiratory failure. It is now know that curare competes with acetylcholine molecules for receptors at neuromuscular junctions.

Since Waterton's work, many other chemicals have been discovered that also affect the nervous system, two of which are anatoxin and saxotoxin.
Anatoxin affects synapses. Its molecules are very similar to those of acetylocholine but they are not broken down by the enzyme, acetylcholinesterase. Saxotoxin is quite different and blocks the sodium channel proteins in nerve axons.

a During synaptic transmission, acetylcholine is released from the presynaptic neurone into the synaptic cleft.

i Describe how an action potential brings about the release of acetylcholine into the synaptic cleft.

....

....

....

.... *(2)*

ii Describe how this acetylcholine may cause a new action potential to develop in the postsynaptic nerve cell.

....

....

....

.... *(3)*

b Explain how injection of arrow poison into the blood stream causes death by respiratory failure.

....

....

....

....

....

.... (3)

c i Expain how acetylcholinesterase is important in the functioning of a synapse.

....

....

....

....

.... (2)

ii Suggest an explanation for the fact that one of the symptoms of anatoxin poisoning is the excessive production of tears.

....

....

....

....

....

....

.... (3)

d Explain how saxotoxin may cause paralysis.

....

....

....

....

.... *(2)*

14 Read the following passage.

A few deaths have been due to drinking too much water. This occurred either because people thought that water was an antidote to *Ecstasy* (it is only an antidote to dehydration) or because the drug induced repetitive behaviour. Under the influence of *Ecstasy*, people have been known to drink 20 litres of fluid and smoke 100 cigarettes within three hours. Drinking pure water to replace the fluid lost in sweating is dangerous because it does not put back the sodium ions and so dilutes the blood. This makes the cells of the body swell up, which is particularly lethal in the brain, which can expand and be crushed against the skull. The centres in the brain that regulate breathing and the heart can be irreversibly damaged, and the person dies. *Ecstasy* increases the risk of brain damage by triggering the release of antidiuretic hormone (ADH).

(Reproduced by permission of *The Guardian*)

a Describe the mechanisms which lead to copious sweating by a person dancing in a nightclub.

....

....

....

....

....

....

....

....

.... *(5)*

b Explain in terms of water potential why the drinking of large amounts of pure water after copious sweating may lead to swelling of cells in the brain.

....

....

....

....

....

....

....

....

....

.... *(5)*

c Explain how *Ecstasy* increases the risk of brain damage by triggering the release of antidiuretic hormone (ADH).

....

....

....

....

....

....

....

....

.... *(4)*

15 Read the following passage.

Diabetes

Diabetes mellitus is a group of disorders that all lead to an increase in blood glucose concentration (hyperglycaemia). The two major types of diabetes mellitus are type I and type II. In type I diabetes there is a deficiency of insulin. Type I diabetes is also called insulin-dependent diabetes mellitus because regular injections of insulin are essential. It most commonly develops in people younger than age twenty.

Type II diabetes most often occurs in people who are over forty and overweight. Clinical symptoms may be mild, and the high glucose concentrations in the blood can often be controlled by diet and exercise. Some type II diabetics secrete low amounts of insulin but others have a sufficient amount or even a surplus of insulin in the blood. For these people, diabetes arises not from a shortage of insulin but because target cells become less responsive to it. Type II diabetes is therefore called non-insulin-dependent diabetes mellitus.

a Describe how blood glucose concentration is controlled by hormones in an individual who is **not** affected by diabetes.

....

....

....

....

....

....

....

....

....

....

....

....

....

....

.... 6

b Suggest how diet and exercise can maintain low glucose concentrations in the blood of type II diabetics.

....

....

....

....

....

....

....

.... *3*

c Glucose starts to appear in the urine when the blood glucose concentration exceeds about 180 mg dm^{-3}. Explain how the kidney normally prevents glucose appearing in the urine.

....

....

....

....

....

....

.... *3*

Section A

1 a Glucose = 6, X = 4, Y = 6 *1*

b Oxidation/dehydration/reduction; decarboxylation/removal of carbon dioxide; Phosphorylation max *2*

c ATP inhibits PFK; Large amounts of ATP means no more ATP produced/reverse argument. *2*

5

2 a ATP *1*

b i Electrons raised to higher level; passed through chain of electron acceptors; Protons/hydrogen ions; from photolysis/water max *2*

ii Reduces; glycerate phosphate/GP *2*

5

3 a Principle of two 'peaks' *1*

b i Continuous; since large number of 'groups' *2*

ii Phenotype affected by environment and genotype/inheritance polygenic *1*

4

4 a Larger groups divided into smaller groups *1*

b i Cercopithecus *1*

ii Cercopithidae *1*

c All show some similarities since in the same family; Blue and green monkeys more similar therefore more closely related than red and white; All show some differences since different species. max *2*

5

5 a Both red and blue receptive cones stimulated *1*

b Highest density of cones; most with individual connection to sensory neurones; cones responsible for colour vision; little/no convergence with cones max *2*

c Less curvature leads to less refraction; therefore light focused further back in eye max *2*

d Lens refracts light, so light less likely to be focused on retina; thickness of lens can be altered to focus on objects at different distances *2*

7

6 a Initiate muscle contraction; by allowing actomyosin bridges to form *2*

b i More cross bridges, the greater the force generated; these shorten, pulling filaments over each other, thus generating force *2*

ii Diagram showing full overlap of actin and myosin *1*

5

7 a i Receive nerve impulses from receptors

ii Interpret/process sensory input

iii Have output to muscles/effectors *3*

b i eg. Damage to association area therefore unable to decide which words to use

ii Damage to motor areas on left side of brain which controls right side of body *2*

5

8 a Allele **A** needed to produce enzyme A; which converts precursor to intermediate; needed as substrate to produce purple colour *3*

b Gametes: Ab, ab × aB, ab
F1 genotypes: AaBb, Aabb, aaBb, aabb

Gametes correct; cross shown by e.g. Punnett square; F1 genotypes correct; F1 phenotypes correct, i.e. 3 white : 1 purple *4*

7

9 a i Rats with $\mathbf{W^r}$ allele at disadvantage since need more vitamin K *1*

ii Homozygotes $\mathbf{W^sW^s}$; selected against; $\mathbf{W^rW^s}$ more likely to survive and pass allele on *2*

b Decline in frequency; since heterozygotes no longer at advantage/ homozygotes $\mathbf{W^sW^s}$ now at advantage *2*

c Deaths occur mainly over age of 50; after most people have had children *2*

7

10 **idea** of isolation;
mechanism of isolation;
environments became different;
mutations;
variation;
natural selection/idea;
operated differently under different conditions *until* populations became unable to interbreed *max 4*

4

11 a i B (allow A)

ii C (allow the muscle which is antagonistic to that given in a i) *2*

b i Mark for principle, maximum of one only
Reflex arc consisting of neurones linking receptor to effector through spinal cord/CNS

For detail, to a maximum of three marks:
Afferent/sensory neurone from receptor;
Intermediate neurone/relay/ internuncial;
Efferent/motor neurone to effector/muscle *Max 3*

ii Allows rapid/immediate response/does not require thought/specific advantage of this reflex, eg; prevent further damage *1*

6

12 a i chloroplast

ii thylakoid *2*

b Actual size = image size/magnification

Size = 99 mm/38 000
= 0.0026 mm
1 mm = 1000 µm
= 2.6 µm *2*

c Internal membranes have large surface area for maximum light absorption; contains pigments capable of trapping light energy.
Flat shape = rapid diffusion in/out; contains relevant enzymes; any 2 *2*

6

Section B

13 a i Calcium ions enter presynaptic membrane; Vesicles fuse with membrane and release contents *2*

ii Diffuses across cleft; Fits receptor site on membrane receptors; Brings about opening of sodium channels; Sodium ions enter postsynaptic neurones max *3*

b Binds with receptor molecules at neuromuscular junctions; Acetylcholine will not fit; Does not trigger muscle contraction; Breathing not possible max *3* d Sodium ions cannot enter neurone; so no action potential/nerve impulse produced *2*

15

14 a Increased muscular activity results in increased heat production in muscles; leading to rise in blood temperature; detected by thermoreceptors; in hypothalamus; impulses to sweat glands; reference to peripheral thermoreceptors max *5*

b Blood/body has lost salt/ions; pure water (taken in) has water potential of 0/ no ions/ less negative water potential than sweat lost; therefore blood has less negative; higher water potential; therefore diffusion gradient/water potential gradient between blood and cells; water therefore diffuses into cells/ moves from a higher to lower water potential; moves along concentration gradient/relevant reference to osmosis. max *5*

c Excess water normally lost by kidneys/ as urine; ADH increases permeability; of distal tubule/ collecting ducts; water reabsorbed (into blood); idea that swollen brain cells damaged by e.g. crushing max *5*

15

15 a For principle, maximum of one mark
Process involves insulin and glucagon

For detail, up to a total of 6 marks
Insulin/glucagon secreted by pancrease/islets of Langerhans;
Hormone receptors in membrane (of target cells);
(**stimulates**) conversion of glucose to glycogen/glycogenesis;
activates/involves enzymes;
stimulates uptake by cells;
conversion of glucose to lipid/protein;
glycagon stimulates conversion of glycogen to glucose/glycogenolysis;
glucagon stimulates conversion of lipid/protein to glucose/glyconeogenesis.
max 6

b Feed on polysaccharides/named example (not cellulose);
slower digestion therefore no surge in blood sugar level;
exercise - increased respiration/BMR.
3

c All glucose is reabsorbed;
involves active transport;
at proximal convoluted tubule. *3*

12